HSU PRESS

ideaFest Journal

Volume 5 | 2021

The Interdisciplinary Journal of
Creative Works and Research
from Humboldt State University

Humboldt State University Press

Humboldt State University Library
1 Harpst St.
Arcata, CA 95524-8299

digitalcommons.humboldt.edu

ISBN: 978-1-947112-63-6

Managing Editors: Aaron Laughlin and Kimberly Sisu
Layout and Design by Aaron Laughlin

Humboldt State University Press publishes high-quality scholarly, intellectual, and creative works by or in support of our campus community. The press supports the HSU mission to improve the human condition and our environment by promoting understanding of social, economic, and environmental issues.

ideaFest Contents Journal

Letter From the Editor

To begin, I would first like to thank all of the wonderful authors and reviewers who contributed to the journal, along with my fantastic co-editor Kim Sisu whose hard work, insights and support were instrumental to this edition. I would also like to extend a special thank you to Kyle Morgan. Kyle—it truly cannot be overstated how invaluable your consistently thoughtful, kind, and intuitive guidance has been, not only in making this journal a reality, but in supporting me during my incredible opportunity to work at HSU Press. It has been a privilege working with you all.

Being the managing editor for *ideaFest Journal* has been quite the journey. As is the case with any good journey, it has been a process of transformation, growth, and learning. I came to the journal as a brand new graduate student interested in the world of editing and publishing, and in less than a year, not only was I suddenly managing a peer-reviewed journal, but the world found itself in the midst of the COVID-19 pandemic along with imperative social unrest and change. During this whirlwind of a time, I have worked closely and rigorously with authors and peer-reviewers, including students, alumni, faculty, staff and community members to hone their research and creative work and present it in a peer-reviewed journal. Managing the process of peer-review, including conducting multiple rounds of edits for every submission we receive, can be a daunting one—especially during a pandemic. I am proud of the collaborations I have participated in and for guiding these incredible authors from start to finish. This work has not only helped me to become a more confident and capable professional and community member, but it has helped authors to develop and share their thoughtful and important scholarly and creative work.

When I applied for a position at HSU Press I aspired to help authors, like myself, to bring their work into the world and share it—in order to assist in facilitating a platform for diverse voices and to promote understanding. With an undergraduate degree in Religious Studies and as a graduate student in English, I have always valued working closely with, and maintaining sensitivity towards, folks from a diverse range of cultural and social backgrounds. It is paramount for the health of our communities and ourselves to support one another in order to further a vision of inclusivity and promote an understanding of differences. I believe *ideaFest Journal*, as an interdisciplinary journal that showcases a wide-range of perspectives and approaches, has proven to be one contribution to that vision.

In my own scholarship, I am primarily interested in the study of mythology in its many forms, which, for me, really amounts to an attempt to understand the power of a good story. This is what publishing is to me; whether it is research journals, novels, poems, music, visual art, etc.—it is about sharing stories, stories that have the potential to shape our realities. Promoting meaningful and ethical change in the world, to create relationships of health and balance for all of its interconnected inhabitants, starts with our stories. I am immensely grateful for being able to play a role in shaping and sharing this work. I would like to encourage authors—and the community of folks behind the scenes in publishing—to continue to strive towards collaboration, compassion, and balance, to share valuable knowledge and help create better stories for us all.

Aaron Laughlin
Managing Editor *ideaFest Journal*

Acknowledgements

Thank you to **Humboldt State University** for being a place of innovative and hands-on education. Student research and authorship requires a dedicated and supportive community, and I feel so fortunate to be in a place of so many like-minded educators.

Thank you to our two-time Managing Editor **Aaron Laughlin**. I did not think it possible to raise the level of this journal, but you have astounded me. Your editorial acumen, facilitation of rigorous peer review, and intensive work with authors have improved the quality of the articles. Your outreach and supportive mentorship have brought in a higher number of articles from a broader scope of fields. Your drive to include voices from a wider range of identities has made this publication more comprehensive and representative. Your innovative, clean and professional design is the envy of academic presses everywhere. Thank you does not even begin to convey my full gratitude.

Thank you to **Kim Sisu** for providing expert editorial work, author support, and publication facilitation. The future of *ideaFest Journal* is in most capable hands.

Thank you to peer reviewers **Marie Antoine, Kathleen Andereck, Frank Fogarty, Nicholas Perdue, Lucy Kerhoulas, Kaitlin Reed, Kathy Thornhill, Catalina Cuellar-Gempeler,** and **Kimberley Stelter.** Thank you for supporting the idea of student research and publication at HSU. You are all true treasures!

Thank you to **David Stacey** and his **ENGLISH 406 students** for providing peer review analyses on all submitted manuscripts. The students provided insightful reviews in the name of campus research, and themselves received a unique hands-on experience to a core skill of scholarly communications. Dr. Stacey, may you never retire.

Thank you to **Cyril Oberlander** for your vision and dedication to publish original student research. Information literacy is more than just finding information. It is about what you do with that information, how you shape it into knowledge, and share it with the world. Your dedicated work empowers students, their voices, and their lives. Thank you for making all this possible.

Thank you to **Janelle Adsit** for overseeing the first volume and laying the foundation upon which all future volumes stand. How can one person be responsible for so much innovative work?

Thank you to **Kacie Flynn**, **Susan Brater**, and **Sponsored Programs** for your support for the journal. Your commitment to student research and authorship has transformed this campus.

And a special thank you to all those who put yourself on the line by submitting an article. Whether published or not, you are all risk takers who strive to create change by sharing your voice with the world. Those who were not chosen for this issue, I expect to see your submissions for the next publication. Success does not come easy, but once achieved, it is something to cherish.

Kyle Morgan
Scholarly Communications and Digital Scholarship Librarian
Humboldt State University Press

A Review of the Contributions of Lichen to Building Soil

Eli Kallison (Humboldt State University)

Keywords: Lichen, Ecology, Soil Building, Weather, Succession

Introduction

Lichens have long been implicated as having an important role in the formation of soil and as primary successional pioneers (Plitt, 1927). In fact, lichens are thought to be essential in their role as pioneers, working to make otherwise uninhabitable ground hospitable to tracheophytes and other plants (Cooper, 1953). They are often the first line of biological invasion on bare rock and recent lava flows. This transformation to a landscape habitable by tracheophytes is done primarily through the process of soil formation by lichens. In addition to forming soil, lichens act as intersystem nutrient sinks, contributing to the enrichment of a landscape with minerals and nutrients. Despite lichens' ecologically significant roles, they remain understudied and enigmatic, particularly in comparison to other organisms with similarly important ecological functions. That being the case, this review is intended to organize and present the details that *do* exist in the literature about specific mechanisms and processes relating to soil formation by lichens. Topics covered here include their role in biodeterioration (both physical and chemical), particle entrapment, decomposition, and lichens' broader role as pioneers in the context of soil formation.

Physical Biodeterioration

Perhaps the most intuitive contribution by lichens to soil formation is through physical biodeterioration. Biodeterioration is the process of changing the properties of a material through the action of biological agents—in this case, lichens. Many lichens have rhizines, which are multicellular root-like structures arising mostly from the lower surface of the lichen thallus, or the lichen "body." Rhizines are used to anchor the lichen to a substrate, and they do not play any direct role in nutrient uptake. In the process of anchoring the lichen thallus to its substrate, rhizines probe and explore surfaces of rocks. As lichens go through their normal process of wetting and drying, their rhizines expand and contract with the moisture fluctuation (Adamo and Violante, 2000). This expansion and contraction of their rhizines act like chisels and contribute to the disaggregation and fragmentation of the rock surface below the lichen thallus. Because lichen thalli can expand and contract by up to 300% of their dry weight during wetting, this chisel action can act quite quickly (Creveld, 1981). The disintegrated rock is broken up into smaller fragments, which can either be incorporated into soil or taken up by the lichen thallus (details of which are discussed later in the section about particle entrapment).

Recent lava flows from volcanoes were studied in Mt. Etna in Sicily, Italy and Mt. Vesuvius in Campania, Italy. These lava flows showed that lichens' rhizines do indeed physically weather rocks. At both the Etna and Vesuvius sites, the authors took Scanning Electron Microscope (SEM) and Energy-dispersive X-ray spectroscopy (EDS) images to analyze rock layers beneath lichen thalli. In both cases, the rock substrate was filled with micro-cracks and pores, which were filled with lichen material, suggesting the chisel-like action created the pores in the rocks (Vingiani et al., 2013). This physical weathering by the lichens' rhizines also induces and accelerates other forms of physical and chemical weathering, making the surface more susceptible to erosion and other forms of mechanical weathering (Chen et al., 2000). One example of this is in cold areas, such as the Arctic, where lichens often dominate the biomass of the area. These regions undergo regular freezing-thawing cycles throughout much of the year. Because the rhizines of lichens penetrate relatively deeply in their substrate, water is able to travel down these channels, collecting deep in the rock's surface where it is more difficult to evaporate. Here the water in the perforations will freeze and thaw, and the ice itself will act as an ice-wedge which can chip away at rocks in tandem with the lichen thallus and the rhizines (Creveld, 1981).

Chemical Biodeterioration

While lichens do play a role in the physical weathering of their substrate, they are even more effective agents of chemical weathering (Jackson, 2015). Oxalic acid, a byproduct of lichen metabolism, is secreted by the mycobiont, or fungal component of a lichen, and is thought to be one of the most active agents of chemical alteration of rocks (Adamo and Violante, 2000). Recently, the magnitude of the effect of both oxalic acid and succinic acid on the deterioration of granite in northeast China was studied. To do this, different concentrations of both chemicals were placed on granite. Both oxalic acid and succinic acid induced the weathering of granite by promoting the release of various ions from the stone and increasing the surface's solubility. Oxalic acid had a stronger effect than succinic acid, and the ions that it released included Na+, K+, Al3+, Fe3+, Mg2+, Mn2+, Ca2+, and SiO2− (Song et al., 2019). While the effects of oxalic acid and succinic acid are significant, they are not the only chemicals secreted by lichens that work to weather rocks.

Carbonic acid has also been shown to increase the rates of substrate weathering. Carbonic acid is generated by respiratory CO_2 reacting with water held by lichen thalli. Carbonic acid increases solubilization processes by lowering the pH of the thallus and surrounding environment (Chen et al., 2000). The lowered pH enhances the dissolution process of the substrate. Carbonic acid and H^+ ions from other organic acids were shown to increase leaching of many metal ions from rocks (Jackson and Keller, 1970). Some researchers even argue that substrate dissolution by respiratory CO_2 may be the most important biogenic weathering mechanism on carbonate substrates (Weber et al., 2011).

In addition to the normal biological byproducts that primarily come from the mycobiont of the lichen thallus, the photobiont also contributes to chemical weathering. Specifically, large biological polymers secreted by the photobiont have been implicated in the chemical weathering of rocks. Instead of directly weathering surfaces, these biological polymers, acidic mucopolysaccharides, act as secondary weathering agents and have been shown to increase dissolution rates of mineral substrates in lab experiments. It is likely that the carboxyl groups in these acidic mucopolysaccharides interact with metals in rocks, thereby increasing the mobility of the metals for easier chelation by the lichen (Barker and Banfield, 1996). These high molecular weight polysaccharides also seem to increase chemical weathering by affecting moisture retention on the surface that can increase disintegration (Banfield et al., 1999).

Before the above modes of weathering were discovered, many researchers in the early and mid-1900's thought that a group of weak, insoluble acids called depsides and depsidones produced by lichens were major players in the chemical degradation of rock substrates (Schatz, 1963). This seemed to be a solid line of reasoning, as lichens produce a unique set of secondary compounds. However, since then, it seems to be the general consensus of researchers that most of these acids play only a very minor role, if any, in the chemical weathering of rock. This is due in part to how weak the acids are, but it is primarily a result of these acids being insoluble in water (Culberson, 1970). The only notable exceptions to this are a number of lichen acids that are slightly soluble in water and therefore available to chelate metal from rocks (Williams and Rudolph, 1974). Even with these slightly soluble acids, however, there is no direct evidence to show that they have any significant and direct effect on weathering of substrates.

A cumulative effect of all the above mentioned modes of chemical weathering leads to long-term dissolution of rocks and other hard substrates into soil. The intensity of disintegration as a result of these characteristics is directly related to physical and chemical properties of the rocks that lichens are deteriorating. Important properties that affect the rate and intensity of degradation are compactness, hardness, lamination, or preexisting surface alteration of the rock. Hard, laminated rocks such as granite are difficult for water and rhizines to penetrate while soft, loose rocks such as limestone are easily broken apart and readily dissolve in weak acids. These factors all affect the accessibility of minerals that lichen need to mobilize in order to degrade rock and convert it to soil. The nature of the lichen thallus is also important, as some lichens secrete more of the chemical weathering elements than others (Adamo and Violante, 2000).

Decomposition

Lichens contribute to soil formation by decomposition as well. There are two primary types of decomposition that contribute to the formation of soil: in-situ decomposition and litterfall. In litterfall, lichens fall to the ground where their thalli are decomposed and incorporated into the soil beneath their epiphytic habitat or downstream to where they are carried. In-situ decomposition is when a lichen dies and decomposes where it initially grew. In this case, like a lichen that has colonized a recent lava flow, the thallus may provide the only source of organic material for decomposition. This mode of soil formation is particularly important, ecologically, for a pioneer ecosystem in which not many other organisms can live.

The amount of organic material that lichens can contribute to soil through decomposition is significant. In California it was found that if canopy cover is 100%, then the amount of thallus tissue decomposing and falling to the ground is 1,020 kilograms per hectare (Knops et al., 1996). In addition to contributing organic matter for decomposition, lichen thalli also slow the rate of decomposition of other litterfall. When lichens and *Quercus* leaves decomposed together, leaves decomposed more thoroughly and more slowly than if they were left to compose on their own (Knops et al., 1996). This effect was particularly pronounced in regards to the retention of nitrogen in the soil. Ultimately, lichens not only contribute a significant amount of organic matter themselves, but they also make the products of decomposition of other matter more valuable to the immediate ecosystem.

Additionally, lichens leach chemicals that affect the decomposition process of other organic matter. Laccases and other chemicals that are important in lichen physiology are also likely to play a significant role in humification (Beckett et al., 2013). Humification is the natural process of transforming organic matter into humic substances that are essential biotic components of soil. Part of this process is the production of a number of enzymes including phenol oxidases, peroxidases, and cellulases that are directly involved in organic matter turnover in free-living saprophytic fungi (Laufer et al., 2009; Liers et al., 2011; Yagüe and Estévez, 1988). These same enzymes are also leached by lichenized fungi, and they most likely function in the same way as those leached by their free-living counterparts. Overall, lichens contribute to soil formation through decomposition in a number of ways, including contributing their thalli as biomass to be decomposed, affecting decomposition rates of other organic matter, and by altering the process of decomposition by leaching enzymes involved in decomposition.

Particle Entrapment

In addition to contributing to organic matter through decomposition, lichens also act as a net for particles in the air and from the substrate that eventually get cycled into the ecosystem as usable matter. As lichens break down rocks, they can incorporate minerals from rocks they are deteriorating into their thallus. This phenomenon of capturing particles and incorporating it into the soil is known as particle entrapment. While particle entrapment occurs as a transfer of particulate matter from rock to lichen thallus, it also occurs as entrapment from the air to lichen thallus. This, along with gaseous uptake of various elements, is one of the primary ways lichens contribute to intersystem inputs via soil building and nutrient enrichment. The different modes of particle entrapment (via air versus substrate) are expressed differently and can be measured by comparing mineral concentrations found in the upper versus the lower portion of the lichen thallus. Mineral compositions of the upper and lower cortex of lichen thalli vary quite drastically: the lower cortex particles match elements commonly found in the substrate, while the upper thallus of lichen have a very different composition that likely came from the air (Clark et al., 2001).

Particle entrapment on the upper cortex of lichens was studied in Italy using a number of species native to the area. Mineral composition between coastal and inland lichen was compared. Lichens living near the Mediterranean Sea had higher concentrations of sodium, potassium, and calcium than normally found in their counterparts farther inland (Vingiani et al., 2013). This increase in nutrient concentration is attributed to the wind-blown sea salts off the Mediterranean being incorporated into the lichen thalli. In fact, the composition of sea-salt particles from the Mediterranean matched those being incorporated into the lichen. The bulk of sea-salt crystals in the Mediterranean Sea are composed of halite (the mineral form of NaCl) and Mg-K sulphate (Benitez-Nelson, 2006). These crystals alone, then, could explain the source of the majority of the nutrient enrichment seen in the seaside lichens and makes a strong case for this being a result of particle entrapment from the air.

On the other side, particle entrapment by lichens' lower thallus comes from their substrate (Banfield et al., 1999). Particle uptake by the lichen thallus is often the result of chemical and physical weathering of their substrate, as discussed in a previous section. The action of the freeze-thaw cycles, chiseling of the rhizines, and chemical weathering via the byproducts of lichen metabolic processes act to free up minerals from their substrates that are then incorporated into the lichen thallus. In one example of this, many lichens are considered "pruinose," which refers to a white dusting of calcium oxalate on the surface of the lichen cortex and throughout the thallus. This trait is considered diagnostic for many lichen genera, and is caused by lichens breaking down rocks that contain calcium carbonate, such as limestone, and incorporating the products into their thalli (Heiđmarsson and Heidmarsson, 1996).

Researchers have even captured the process of mineral uptake from lichens' substrate through microscopy (Barker and Banfield, 1996). As lichens incorporate minerals into their thallus, the minerals themselves are transformed. Mineral transformation is the process of altering the physical and/or chemical structure of minerals. The exact mechanisms of this process, however, are unknown because they were occurring at a scale beyond the resolution of the microscope used in the study (Barker and Banfield, 1996). So, not only do lichens uptake particulate matter in the form of elements from their substrate, they are also potentially converting it into different forms, which might be more usable by living organisms. This phenomenon of nutrient transformation was also observed in Ontario, Canada. Here, lichens were shown to effectively weather their rock substrate Feldspar, which is a rock-forming tectosilicate group of minerals that make up almost half of the Earth's continental crust. In the process of weathering the stone, lichens transformed the elements from Feldspar to clay minerals and they did so at a much higher rate than bryophytes, which was the other study system (Jackson, 2015). This mineral transformation is one of the key reasons why lichens are so important in ecosystems beyond simple soil building. Lichens are able to transform substances and nutrients from unusable consolidated forms to nutritious bioavailable forms, often in the form of chelated metals (Vingiani et al., 2013).

Ecological Succession

Although all of the evidence above is compelling on its own, taken together they form a clear picture of how ecologically important lichens are as pioneer species that enable succession by "higher plants." To summarize, lichens play a role in building soil via a number of mechanisms including physical and chemical biodeterioration, decomposition (litterfall and in-situ), and particle entrapment. We know, too, that in addition to creating soil, they are able to make nutrients more bioavailable, through the chelation and breakdown of metals in rocks, for example. With all this in mind, it is easy to see how lichens enable succession.

In terms of what succession actually looks like, here is an example from Charles Plitt in *The Bryologist*: bare rock is colonized by microlichens who are nutritionally supported by the air and use the rock simply as a substrate. The microlichens begin to break down the rock ever so slightly, and when they die they decay and leave behind a thin organic layer. This layer is substantial enough that foliose lichens can dig their rhizines in, and they start to physically and chemically weather the rock even further. After a few successive generations of these larger lichens, there is a substantial layer of humus laid down through in-situ decomposition, which becomes suitable for colonization by bryophytes. Bryophytes go through a few successional stages and live for quite a while until their decay and breakdown of the substrate create even more soil, allowing for small herbaceous plants to thrive (Plitt, 1927). By this time, the ecosystem is starting to resemble a more typical terrestrial system, and small shrubs start to pop up and they are eventually replaced by larger, longer-lived tracheophytes. In many cases, lichens and bryophytes will start to grow epiphytically on the larger tracheophytes such as trees. Although this is an idealized, typical model of succession from lichens to tracheophytes, the way this happens varies with the specific environmental conditions of a location. Differences in lichen composition, rate of succession, and whether succession occurs at all, is dependent on many variables including altitude, temperature, moisture, sunlight, and more. In deserts, succession will be much different than in alpine environments. And in some ecosystems, such as the harsh Arctic, succession beyond lichens and some bryophytes may not occur at all unless induced by global climate change (Nascimbene et al., 2017).

That said, there are some scientists who claim that lichens' role in ecological succession is greatly exaggerated, particularly on recent lava flows (Williams and Rudolph, 1974). For example, after the explosion of a volcano (which is what would render new lava flows) there is enough dust and debris from the explosion that lichens are unnecessary for the creation of soil. Surveys taken after a recent explosion found enough debris to be considered soil hidden even deep in the nooks and crannies of lava rock. This soil is hypothesized to be suitable for plant roots and bryophytes (Williams and Rudolph, 1974). Overall, though, particularly in more modern literature, there seems to be a general consensus that lichens *do* play a crucial role in soil formation and succession.

Acknowledgements

Marie E. Antoine who is the foundation and motivation for my knowledge about lichens.

References

Adamo, P., and P. Violante. 2000. Weathering of rocks and neogenesis of minerals associated with lichen activity. *Applied Clay Science* 16: 229–256.

Banfield, J. F., W. W. Barker, S. A. Welch, and A. Taunton. 1999. Biological impact on mineral dissolution: Application of the lichen model to understanding mineral weathering in the rhizosphere. *Proceedings of the National Academy of Sciences* 96: 3404–3411.

Barker, W. W., and J. F. Banfield. 1996. Biologically versus inorganically mediated weathering reactions: relationships between minerals and extracellular microbial polymers in lithobiontic communities. *Chemical Geology* 132: 55–69.

Beckett, R. P., A. G. Zavarzina, and C. Liers. 2013. Oxidoreductases and cellulases in lichens: Possible roles in lichen biology and soil organic matter turnover. *Fungal Biology* 117: 431–438.

Benitez-Nelson, C. 2006. Book Review | Chemical Oceanography (Third Edition). *Oceanography* 19: 153–154.

Chen, J., H.-P. Blume, and L. Beyer. 2000. Weathering of rocks induced by lichen colonization — a review. *CATENA* 39: 121–146.

Clark, B. M., L. L. S. Clair, N. F. Mangelson, L. B. Rees, P. G. Grant, and G. S. Bench. 2001. Characterization of mycobiont adaptations in the foliose lichen Xanthoparmelia chlorochroa (Parmeliaceae). *American Journal of Botany* 88: 1742–1749.

Cooper, R. 1953. The Role of Lichens in Soil Formation and Plant Succession. *Ecology* 34: 805–807.

Creveld, Marijke. 1981. Epilithic lichen communities in the alpine zone of southern Norway. J. Cramer, Vaduz Liechtenstein.

Culberson, W. L. 1970. Chemosystematics and Ecology of Lichen-Forming Fungi. *Annual Review of Ecology & Systematics* 1: 153–170.

Heiđmarsson, S., and S. Heidmarsson. 1996. Pruina as a Taxonomic Character in the Lichen Genus Dermatocarpon. *The Bryologist* 99: 315.

Jackson, T. A. 2015. Weathering, secondary mineral genesis, and soil formation caused by lichens and mosses growing on granitic gneiss in a boreal forest environment. *Geoderma* 251–252: 78–91.

Jackson, T. A., and W. D. Keller. 1970. A comparative study of the role of lichens and 'inorganic' processes in the chemical weathering of Recent Hawaiian lava flows. *American Journal of Science* 269: 446–466.

Knops, J. M. H., T. H. Nash, and W. H. Schlesinger. 1996. The influence of epephytic lichens on the nutrient cycling of an oak woodland. *Ecological Monographs; Durham* 66: 159.

Laufer, Z., R. P. Beckett, F. V. Minibayeva, S. Lüthje, and M. Böttger. 2009. Diversity of laccases from lichens in suborder Peltigerineae. *The Bryologist* 112: 418–426.

Liers, C., R. Ullrich, M. Hofrichter, F. V. Minibayeva, and R. P. Beckett. 2011. A heme peroxidase of the ascomyceteous lichen Leptogium saturninum oxidizes high-redox potential substrates. *Fungal Genetics and Biology* 48: 1139–1145.

Nascimbene, J., H. Mayrhofer, M. Dainese, and P. O. Bilovitz. 2017. Assembly patterns of soil-dwelling lichens after glacier retreat in the European Alps. *Journal of Biogeography* 44: 1393–1404.

Plitt, C. C. 1927. Succession in Lichens. *The Bryologist* 30: 1–4.

Schatz, A. 1963. Chelation in Nutrition, Soil Microorganisms and Soil Chelation. The Pedogenic Action of Lichens and Lichen Acids. *Journal of Agricultural and Food Chemistry* 11: 112–118.

Song, J. F., J. X. Ru, X. P. Liu, and X. Y. Cui. 2019. Oxalic Acid and Succinic Acid Mediate the Weathering Process of Granite in the Cold-Temperate Forest Regions of Northeast China. *Eurasian Soil Science* 52: 903–915.

Vingiani, S., F. Terribile, and P. Adamo. 2013. Weathering and particle entrapment at the rock–lichen interface in Italian volcanic environments. *Geoderma* 207–208: 244–255.

Weber, B., C. Scherr, F. Bicker, T. Friedl, and B. Büdel. 2011. Respiration-induced weathering patterns of two endolithically growing lichens. *Geobiology* 9: 34–43.

Williams, M. E., and E. D. Rudolph. 1974. The Role of Lichens and Associated Fungi in the Chemical Weathering of Rock. *Mycologia* 66: 648–660.

Yagüe, E., and M. P. Estévez. 1988. Regulation of β-1,4-Glucanase and β-Glucosidase Production by Glucose in Evernia prunastri. *Journal of Plant Physiology* 133: 539–544.

Effects of Redwood Bark and Leaf Leachate on Different Lichen Populations Found Within the Redwood Forest of Arcata, CA

Courtney Michelle Harris (Humboldt State University), Phoebe Hughes (Humboldt State University)

Keywords: Sequoia semperviren, Picea stichensis, Alnus rubra, redwood chemistry, leaf leachate, bark leachate, Flavoparmelia sp., Hypogymnia sp., Parmelia sp., Parmotrema sp., Ramalina sp., Sphaerophorus sp., Usnea sp., lichen, lichens

Introduction

The redwood-dominated coastal forest of Northern California is a unique environment that is home to several endemic, and many closely associated, tree and lichen species. Associated tree species such as the red alder, *Alnus rubra,* Bong., and Sitka spruce, *Picea sitchensis* (Bong.) Carriere, can often be found within the same forest habitat as the coast redwood, *Sequoia sempervirens* (Lamb. ex D. Don) Endl. Although these associated trees share the same space and common epiphytes, the lichens that are present on the branches and in the canopy of these associated trees are much less abundant on the redwoods, if they appear at all.

Lichens exist as epiphytes with an ecologically obligate mutualistic symbiosis that is shared between a fungal partner, the mycobiont, and an algal partner, the photobiont. Their photobiont can be either green algae, known as a chlorobiont, and/or cyanobacteria, a cyanobiont. These partners provide the lichen with sugars, or, in the case of the chlorobiont, both sugars and fixed nitrogen. The structure of lichens, as dictated by their mycobiont, can vary in form from crustose to foliose, fruticose, or pendulous, and often take on various intermediate forms, allowing for their survival in almost any environment. These organisms are poikilohydric which means they experience varying concentrations of water in their body, the thallus, as water is absorbed passively through fluctuations in concentration gradients on their surfaces. As poikilohydric organisms, moisture is required for the transport of nutrients and water into the lichen and, accordingly, most lichens are found in areas that can meet their hydration as well as their climatic and nutritional needs. This poikilohydric nature renders lichens vulnerable to water-soluble environmental toxins, as they are not able to filter out or discern between harmful and beneficial molecules.

About thirty-four percent of the water that is annually available in the coastal redwood forest comes directly from the fog drip collected by the foliage of the redwoods, and contributes to up to sixty-six percent of its water during the hottest parts of the year (Dawson, 1998). The rainfall in redwood forests is less than 1 meter per year, so the redwoods use the limited rainfall in conjunction with the trapped coastal fog to drip for hydration. It would make sense, then, for lichens to take advantage of the redwood canopies, as there is plenty of moisture and sunlight; however, even when an associated tree harboring a diverse community of lichens extends to touch the redwood's branches, most of the lichen communities will not transfer over and establish on the redwoods as they would on other trees. Based on the observations of lichen species within the Douglas fir-dominated forests, the limited presence of lichens on non-redwoods has been hypothesized to have been due to dispersal limitations (Sillett et at., 2000). However, later observations in the redwood-dominated forest showed that even when a lichen was given ample redwood substrate, lichens (especially cyanolichens) were still resistant to habitation on the available redwood's surface (Williams et al., 2007). Another hypothesis for the cause of the reduced presence of lichens on redwood substrate might be related to the low-light tolerance of redwoods, which could inhibit the growth of other epiphytic species. However, lichens are adapted to become light-saturated at very low light levels, conditions that would inhibit other species, affording lichens the advantage.

If redwoods exhibit ample open substrate space for lichen colonization, and the lichens are not limited by dispersal methods or low-light conditions, then it stands that there must be another explanation as to why lichens do not colonize on redwoods. Redwood trees are well known to have a unique chemical composition in their foliage as well as their bark, and it is thought that these compounds may be the limiting factor to many other genera of fungi, plants, and animals taking up residence within the redwood forest. The bark of redwood trees exhibit response mechanisms against herbivory such as the swelling of polyphenolic parenchyma, an increase in cell wall lignification, and traumatic resin duct development within the stems of the redwood (Hudgins, 2004). Polyphenolic compounds have been identified to retard the growth of pathogenic fungi that might try to inhabit the tree (Hall, 1985). This method of protection works to deter herbivores as well, but redwoods also exhibit other forms of protection. The inhibition of endophytic fungi growth is not limited to the bark; different terpenes found within the foliage of redwoods have also been identified as inhibitors of the growth of some of their endophytic fungal partners (Espinosa-Garcia, 1991). The foliage of redwood trees contains many known and identified terpenes, some of which have varying concentrations depending on the stage of maturation of the tree, suggesting yet another form of herbivory defense (Okamoto et al., 1981).

Given that lichens have ample available substrate, water, and light within the redwood forest, it seems most likely that chemical toxicity is the true limiting factor for lichen presence in the lower strata of redwoods. Lichens are more likely to colonize the bark of their hosts, rather than the foliage, and taking into consideration that lichens absorb nutrients as well as toxins through their hydrated thallus, we hypothesize that it is the chemical composition of the redwood bark that is the main deterrent to lichen colonization on redwood trees, and that these chemicals are leached out of the bark through rainwater passing over the outer surface of the tree. Therefore, if the chemical composition of either the bark or the foliage prove to be detrimental to lichen growth, the observable decrease in lichen coverage when exposed to leachates from different parts of the redwood tree will not be zero. In other words, if there are chemicals within these parts of the tree that are inhibiting the growth of lichens, then there will be a significant negative change in the percent coverage of lichens. This study investigated the effect of redwood leaf leachate and redwood bark leachate on the growth of lichens from redwood forest associated tree species. We hypothesized that the chemical compounds in *S. sempervirens* will negatively affect the health of the lichen populations found on the neighboring associated tree species.

Materials and Methods

To perform this experiment, we watered the specimens of collected sticks, which were covered in representative lichens from the different associated tree species of the redwood forest, with leachates prepared from different parts of the redwood tree: the bark and leaves. The prepared leachates are meant to mimic the natural chemistry that might be found in the water that passes down the trunk of a redwood or cascades down through the needles during rainfall. To set up an experimental area, a metal kitchen rack with four metal wire shelves, measuring 4 feet x 2 feet x 5 feet, was placed in a protected area behind the greenhouse of Humboldt State University. The area was shaded by tall surrounding buildings, protected from excessive wind, and the rack was placed close to a wall, with no direct canopy cover. The rack was adorned with twenty-four metal hooks that were placed at the front and the back of each shelf in three sets of two on each level, forming three columns of four sticks each, with the capability of holding a total of twelve sticks on the rack (see Figure 1). The twelve experimental sticks were then gathered from the floor of Arcata Community Forest the day following a significant windstorm, to ensure the freshness of fallen material. Through casual observation, the main associated tree species within this portion of the redwood forest were determined to be red alder (*Alnus rubra*) and Sitka spruce (*Picea sitchensis*).

Healthy sticks were gathered at random and kept for the experiment based on three main criteria. First it was determined by visual identification if the stick belonged to either a red alder or a Sitka spruce tree. The second criterion was the overall good health of the lichens present on the stick, as assessed by observational analysis. Finally, the third criterion was that the stick be about 1 to 2 inches in diameter and at least 24 inches long, or be able to be cut to that length without damaging the lichen communities on the stick. Macrolichen species were identified using *Macrolichens of the Pacific Northwest* (McCune and Geiser, 1997, Table 1). The first six sticks of each associated tree species that met these criteria were transported to the greenhouse in a 5-gallon bucket and hung from the metal hooks. The different stick species were separated such that each of the three columns on the rack had two Sitka spruce branches and two red alder branches to ensure replication between the three treatment groups. The sticks were labeled with a number 1 through 12 that was written with a permanent marker on plant tape, which was then tied around one end of each stick starting with 1 on the top left and then numbering down.

The initial percent-coverage of lichens on each stick was determined by dividing the stick into four equal quadrants on the front side of the stick and four equal quadrants on the back. Each side of the stick was observed individually as its own whole and treated as a two-dimensional surface, where each of the four equal quadrants represented 25% coverage. Calculated results of both sides were then added together and divided by two to get the total percent coverage of the stick. This is also how final percent coverage was calculated, and the difference between the final and initial percent coverage yielded our observed decrease in lichens. After initial coverage was determined, the sticks received their first treatment.

Each of the three columns received a different treatment and thus were numbered vertically. The far-left column of sticks, numbered 1-4, were the control group. In the middle column, the sticks were numbered 5-8 and given the bark leachate experimental treatment, and in the far-right column, the sticks were numbered 9-12 and received the leaf leachate experimental treatment. The treatments were given in vertical groups to avoid cross-contamination of treatment types due to inevitable dripping.

The three treatments were prepared as follows; the control group was treated with untreated rainwater, which was collected in a 5-gallon bucket in the backyard of HSU's greenhouse. The second treatment, bark leachate, was prepared with approximately 700 grams of dry redwood outer-bark, weighed with a balance, then macerated with gardening shears and saturated with 8 liters of rainwater in its own 5-gallon bucket. The third treatment was composed of approximately 700 grams of redwood foliage, which was cut from the attached lateral woody branches, weighed with a balance, and covered with 8 liters of rainwater in a third 5-gallon bucket. These leachates were prepared once a week, two days prior to the first treatment day of the week. The mass of redwood bark and foliage used was determined by the availability of the sample that could be collected without damaging the redwood tree and the amount of rain water used was determined by how much it would take to completely submerge the redwood bark and leaf samples inside the 5-gallon buckets without depleting our limited rainwater reserves. The treatments were administered every Tuesday, Thursday, and Saturday morning before the hottest part of the day to ensure maximum retention of the treatment solutions given. The pH of each leachate solution was measured after each treatment with the Thermo Scientific Orion Star A-111(c) pH meter, which was calibrated with a pH 3 and pH 7 buffers prior to each use. Each treatment group was given one full liter of leachate per treatment by means of a hand-held spray bottle, which was split amongst the four specimens of each group. To administer the treatments, each stick was carefully removed from the hooks and sprayed until saturated, mimicking heavy rainfall, then placed back on the hooks in its column to drip dry as it would in nature. The treatments were repeated for five weeks.

During the treatment duration, observational data was recorded by means of bi-weekly photos. Twice a week, on days that alternate treatment days, photos were taken of each stick in each treatment group with an Apple iPhone and then uploaded to a google doc. To do this, each stick was carefully removed from its hooks and placed in a flat black-colored tray alongside a 24-inch ruler. Pictures were then taken of the front and the back sides of each of the twelve sticks, paying particularly close attention to areas that may be exhibiting signs of impact from the treatment being administered. The photos at the end were then compared to the photos taken before treatment began, and % coverage decrease was recorded using a quadrat and the method described above.

During this experiment, the pH values of each of the treatments were recorded, and although the pH values that were recorded seemed as if they may provide some valuable information in this experiment, we decided not to use this data. The averages that were recorded were based on the treatments that were collected only in the first three weeks of experimentation. After the first three weeks, data collection was discontinued due to campus closures relating to the COVID-19 outbreak, which began during the latter half of our experiment and caused the loss of access to the Thermo Scientific Orion Star A-111(c) pH meter.

When analyzing the data, a one-way ANOVA test was used to compare the mean decrease in lichen coverage of the three treatment groups, pooling the data from all host sticks regardless of species. A one-way ANOVA test was also used to compare the difference in decrease of lichen coverage between host-stick species that were given the same treatment, and the test was run for each treatment group. We then used the p-values to determine statistical significance.

Results

The bark leachate treatment had no significant effect on the percent decrease of the lichen population on either stick species (p-value= 0.55). The leaf leachate did have a significant effect on the percent decrease of the lichen population on both host stick species (p-value= 0.001). The pH of the bark leachate averaged 3.5, the leaf leachate averaged 5.73, and the control averaged 6.8 (See Table 2, Figures 2 and 3).

Discussion

Our first analysis yielded a graph that showed the percent decrease in lichen coverage on the host sticks in each treatment and compared the three results (Figure 2). The difference in the lichen coverage decrease between the lichens on the Sitka spruce host sticks and the red alder host sticks was not statistically significant in any of the treatment groups, so we pooled the data to make a more concise graph that better summarized the data and helped recognize the total difference between treatments (Figure 3). Although the bark leachate did have a slightly higher percent decrease in lichen coverage than the control, the magnitude of the necrosis was not statistically significant. This may have been due to the use of outer bark, versus the use of inner heartwood, which is known to have a higher concentration of polyphenolic compounds and thus may have rendered different results if we used this wood instead. The use of the inner heartwood could more accurately illustrate the negative impact of the polyphenolic compounds on the lichen's overall health through closer proximity to these compounds.

Redwood outer bark, contrarily, is very fibrous and sloughs off easily, which could deter the establishment of macrolichens and may also suggest a different chemical composition than what is found in the inner heartwood. The treatment of leaf leachate, however, had a drastically more significant effect (Figures 2 and 3). Based on the data collected, it can be inferred that lichen growth on redwoods is not significantly impacted by the polyphenolic compounds in the outer bark but is more likely impacted by the compounds in the foliage leaching out and down the bark's surface through fog accumulation and rain. It is true that the wood has its own set of inhibitory, aromatic compounds, but they are found deep within the heartwood of the tree. One hypothesis we propose, based on this fact, and our findings with this experiment, is that the toxic nature of the foliage is attributed to the polyphenolic compounds of the heartwood being transferred into the sapwood and conducted through the vascular tissue and into the leaves. However, further experimentation on this would be needed to confirm these inferences.

Our results demonstrated that the percent decrease in lichen coverage was virtually the same for both the bark leachate treatment and the control. Therefore, the loss of these lichens could be attributed to initially being moved to a new environment, stress from the constant disturbance of being removed from the rack to be treated and photographed, or other natural and experimental variables. It is also possible that the bark generally has less of these polyphenolic compounds on the outer surface, which is where we obtained our bark samples. Although the bark leachate did not render statistically significant results, one observation was made but not quantified within this treatment group: the foliose *Parmotrema* spp. lichens showed an abundant increase in the production of marginal soredia when photographically compared to its initial photos and the control sticks. Likewise, the fruticose *Usnea* spp. demonstrated an observable increase in apothecia production. When certain lichens experience environmental stresses, it is not uncommon to see an overproduction of reproductive structures as an attempt to reproduce when experiencing conditions that may lead to the death of the lichen. This behavior suggests that the lichens were, in fact, experiencing stress, and over a longer duration it is likely we would have observed an even greater production of these structures and possibly eventual death of the lichen.

Continuing this experiment for a longer duration in the future, as well as using heartwood over bark, may provide better insight into how these lichen communities interact with the chemistry of the redwood trees. Additionally, future research on the chemical composition of water extractions from the leaves and the heartwood may prove to be useful in determining which compounds are present and in what concentration these compounds may become toxic to certain lichen communities. Ultimately, the expansion of this information, when taken into consideration with our experimental findings, could provide better insight into why some lichens are better suited to live in the canopies of redwoods and on neighboring associated trees rather than in the lower strata of redwoods.

Table 1. Identified macrolichen genera found on each stick.

	Flavoparmelia	*Hypogymnia*	*Parmelia*	*Parmotrema*	*Ramalina*	*Sphaerophorus*	*Usnea*
Stick 1 (alder)	x		x	x			x
Stick 2 (alder)		x	x	x	x		x
Stick 3 (Sitka)		x	x	x			x
Stick 4 (Sitka)		x		x	x	x	x
Stick 5 (alder)	x	x	x	x		x	x
Stick 6 (alder)	x			x	x		x
Stick 7 (Sitka)		x		x			x
Stick 8 (Sitka)			x	x	x		x
Stick 9 (alder)			x	x	x		x
Stick 10 (alder)			x	x	x	x	x
Stick 11 (Sitka)			x	x			x
Stick 12 (Sitka)			x	x			x

Table 2. Mean decrease in percent coverage of lichens on red alder and Sitka spruce sticks after treatment with control (rain water), redwood bark leachate, or redwood leaf leachate experimentaltreatments.

	Control Alder	**Control Sitka**	**Bark leachate Alder**	**Bark leachate Sitka**	**Leaf leachate Alder**	**Leaf leachate Sitka**
Coverage Decrease (%) Stick 1	5.0	4.0	5.0	3.0	14.0	10.0
Coverage Decrease (%) Stick 2	4.0	3.0	4.0	7.0	12.0	12.0
Average Coverage Decrease (%)	4.5	3.5	4.5	5.0	13.0	11.0

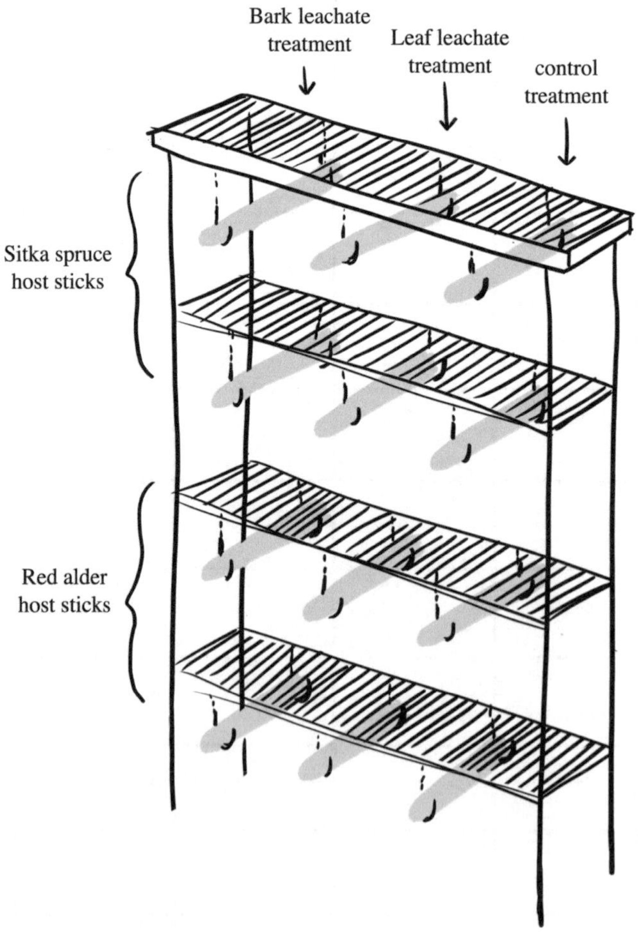

Figure 1. Setup of the kitchen rack, hooks, experimental sticks, and treatment groups.

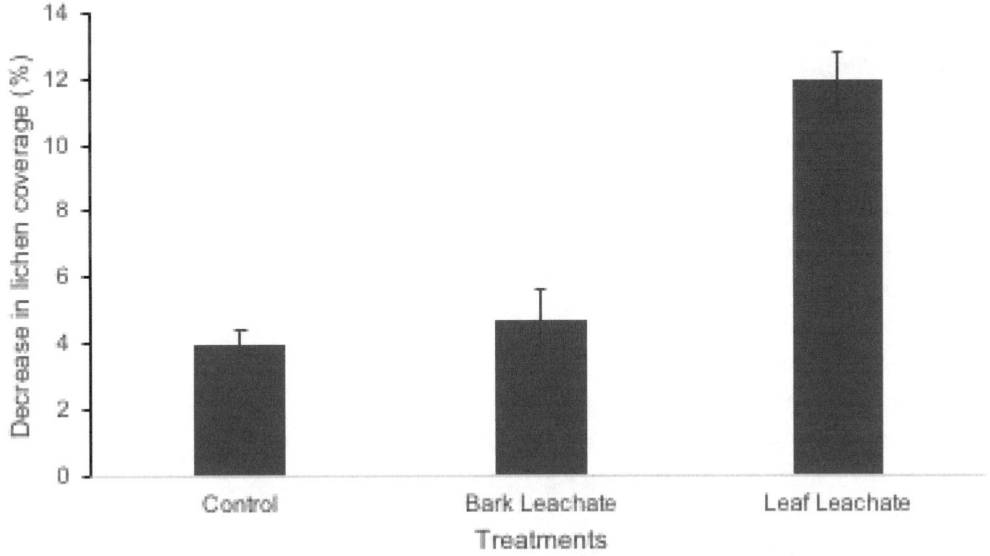

Figure 2. Total percent decrease in lichen coverage between species of host sticks (red alder, Sitka spruce) per treatment group (control rain water, bark leachate, and leaf leachate). Standard error for eachtreatment is shown. The p-value between the control and the bark leachate is >0.05, which isn't significant. The p-value between both control and leaf leachate, and bark and leaf leachate is <0.05, and is therefore statistically significant.

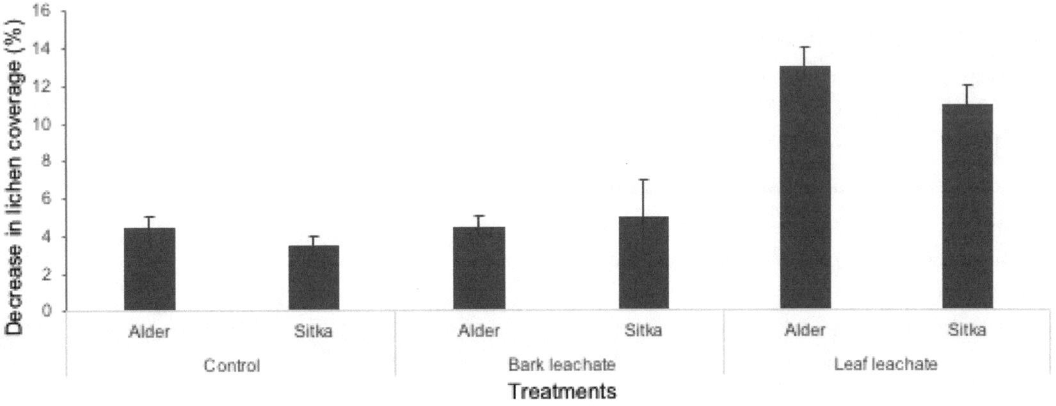

Figure 3. Mean percent decrease in lichen coverage between treatment groups (control rain water, redwood bark leachate, and redwood leaf leachate). The p-values between the host sticks within each ofthe treatment groups was >0.05 and therefore not significant.

Acknowledgements

We would like to thank Marie Antoine for her guidance and assistance with editing this article, and for her inspiring passion in the field of Lichenology.

Citations

Dawson, T. 1998. Fog in the California redwood forest: ecosystem inputs and use by plants. Oecologia 117, 476–485. https://doi.org/10.1007/s004420050683

Espinosa-García, F. J., & Langenheim, J. H. 1991. Effects of sabinene and γ-terpinene from coastal redwood leaves acting singly or in mixtures on the growth of some of their fungus endophytes. Biochemical Systematics and Ecology, 19(8), 643–650. doi:10.1016/0305-1978(91)90080-j

Hall, G. D. 1985. Leaf Monoterpenes of Coast Redwood (Sequoia sempervirens). Ph.D. dissertation. University of California, Santa Cruz.

Hudgins, J. W., E. Christiansen, and V. R. Franceschi. 2004. Induction of anatomically based defense responses in stems of diverse conifers by methyl jasmonate: a phylogenetic perspective. Tree Physiology, 24(3), 251–264. doi:10.1093/treephys/24.3.251

Okamoto, R. A., Ellison, B. O., & Kepner, R. E. 1981. Volatile terpenes in *Sequoia sempervirens* foliage. Changes in composition during maturation. Journal of Agricultural and Food Chemistry, 29(2), 324–326. doi:10.1021/jf00104a026

Sillett, S. C., B. McCune, J. E. Peck & T. Rambo. 2000. Four years of epiphyte colonization in Douglas- fir forest canopies. The Bryologist 103: 661–669.

Woodward, B. D., W. H. Romme, P. H. Evangelista. 2020. Early postfire response of a northern range margin coast redwood forest community. Forest Ecology and Management, 462 https://doi: 10.1016/j.foreco.2020.117966

Combined Use of Data From Avian Surveys Along the Pacific Crest Trail With Biodiversity Repositories to Model Habitat Suitability Throughout Northern California

Holli N. Pruhsmeier (Humboldt State University), Michael C. McGrann (William Jessup University), Jim Graham (Humboldt State University)

Abstract

Models that describe species distributions are valuable in guiding management decisions. We compared and combined two avian datasets during the 2010 breeding season in northern California, USA. These datasets were a large-scale avian diversity survey from McGrann and Furnas (2016; 2018) and combined data from Biological Information Serving Our Nation (BISON) and Global Biodiversity Information Facility (GBIF). Our objective was to compare the utility of these two datasets, that employ separate field protocols, to model habitat use for the Black-headed Grosbeak, Hairy Woodpecker, and Yellow-rumped Warbler, three common forest birds in our study area that occupy distinctive habitat types. We also tested whether combining the datasets together would create a model with greater generality over the study area and determine if the data will create response curves that explain certain relationships between environmental characteristics and species occurrences. We found that fine-scale data along a single, albeit extensive, transect built models that predicted suitability well for the section of trail, but did not predict occurrences well for areas beyond the trail in two of the three species. We also found that data from Biological Information Serving Our Nation (BISON) and Global Biodiversity Information Facility (GBIF) did not have the sampling structure required for finer scale modeling and lacked observations in areas that may be critical for sampling, such as fire-impacted areas. By combining these two datasets, we produced models that captured the range of these species throughout the study area, and we created response curves that explained anticipated habitat associations for each species.

Keywords: BlueSpray, habitat suitability modeling, MaxEnt, passerines, woodpeckers

Introduction

Birds are excellent indicators of environmental change because they rely on plant communities and the overall structure of vegetation to provide food, shelter, and breeding and nesting sites. Their distributions, therefore, will shift as a result of human land use change (Lee et al., 2004) and other factors such as fire or drought (Zimmerman, 1997). In an effort to conserve bird habitat, managers require tools to aid their decision-making processes, including spatial modeling tools (Tingley et al., 2009; Turner et al., 2016). In this study, we use a spatial modeling tool, MaxEnt, in a habitat suitability modeling procedure to select environmental characteristics that determine what habitat types are associated with specific avian species occurrences. It is important to create accurate habitat models to determine areas which may be crucial to establish as nature preserves that anticipate the effects of climate change and human land-use.

Habitat suitability models predict the spatial occurrence and distribution of a species based on measures of habitat suitability (Peterson et al., 2011), with elevation, topography, habitat type, precipitation, and temperature as common measures of habitat suitability (MacArthur, 1965; Hedley & Buckland, 2004; Odion et al., 2010; McGrann & Thorne, 2014; Asner et al., 2015; Kadmon et al., 2016; McGrann & Furnas, 2016). Habitat suitability models are represented as grid-based maps of the spatial distribution of estimated habitat suitability (Kimble, 2016). Habitat models have been created for many avian species, such as the species that we include in this study: Hairy Woodpecker (*Picoides villosus*) (Russell et al., 2007), Yellow-rumped Warbler (*Dendroica coronata*) (Price, 2000), and Black-headed Grosbeak (*Pheucticus melanocephalus*) (St-Louis et al., 2014).

MaxEnt is a spatial modeling tool that has been used in a wide variety of species distribution applications, including the mapping of phenotypic diversity in Hairy Woodpeckers (Klicka et al., 2011), modeling climate-induced shifts in the distribution of Warbler species (Ralston & Kirchman, 2013), and it has been applied in conservation planning by modeling habitat suitability for migratory birds, including the Black-headed Grosbeak (Seavy et al., 2012).

Habitat suitability models use species occurrence data, which are typically geographic locations where the species has been detected in the field using a standardized survey protocol that typically employs some randomized sampling procedure. Ideally, data is collected via these same protocols across the entire study area of interest (Austin & Heyligers, 1989). However, in reality, most published species distribution studies employ very different survey protocols; it is rare to find that data is collected in a standardized manner across more than one study. It is for this reason that most biodiversity databanks and clearinghouses, such as Biological Information Serving Our Nation (BISON) and Global Biodiversity Information Facility (GBIF), offer a collection of existing datasets that may be biased due to the original purpose of the study (Barry & Elith, 2006).

Using a single modeling approach, we compared two datasets of avian species distributions that were collected using varying sampling designs and spatial coverage in northern California. Our analysis compared data collected using a large-scale avian diversity survey along the Pacific Crest National Scenic Trail in northern California (McGrann & Furnas, 2016; Furnas & McGrann, 2018), which we will, henceforth, call the PCT Data, and combined data from Biological Information Serving Our Nation (BISON) and Global Biodiversity Information Facility (GBIF), which we will call the GBIF-BISON Data. The PCT Data represents a study with a standardized methodology for point counts and automated recording units, while the GBIF-BISON Data obtains their data from contributors who have varying methods and data quality.

Our objectives were to test the generality of model predictions derived from each dataset and use these predictions to evaluate three hypotheses. Our hypotheses were: (1) the PCT Data will perform well for modeling the habitat associations of the species for areas near the trail but will decrease in performance with distance from the trail. (2) The GBIF-BISON Data, although composed of observations dispersed throughout the entire study area, will not be detailed enough to transfer to a finer scale analysis. (3) By combining these two datasets, which we will call the Combined Dataset, we can create a model that is fine-tuned to the scale of the analysis but that also generalizes well across our entire study area and creates parsimonious response curves by associating species occurrences with environmental characteristics. In order to test these hypotheses on the generality of model predictions, we felt it was best to compare model predictions for three relatively common and widespread species that also have distinctive niches, habitat associations, and life histories. We reference these species using a six-letter alpha coding system (Pyle & DeSante, 2003): (1) Black-headed Grosbeak (BKHGRO) is a neotropical migrant. The species prefers a complex vegetation composition and structure with a mixture of hardwood and conifer trees (Williams & Koenig, 1980). (2) Yellow-rumped Warbler (YERWAR) is a year-round resident and elevation migrant and considered a habitat generalist, found in all elevations but with a preference for coniferous forests. Both BKHGRO and YERWAR also have close habitat associations with water (Kirkpatrick et al., 2009; Becker, 2013). (3) Also a resident, Hairy woodpeckers (HAIWOO) occur at higher elevations, and although abundant in green forests, they are particularly associated with recently burned areas. In burned forests, there is an abundance of snags, which yield wood-boring insects (Parker et al., 2006), an important food resource (Saab et al., 2019).

Methods

Study Region

We studied a region along the northern California portion of the PCT that extended from Bucks Lake Wilderness (39.907°N, -121.127°W) to the Oregon Border (42.005°N, -122.913°W) (Figure 1). Much of the southern and eastern portion of this section of the PCT lies within the rain shadow of the Cascade Mountains, exhibiting drier conditions. The trail then turns west into the moister Klamath Mountains (McGrann et al., 2014). These conditions create a diverse climate that is predominantly forested, ranging from mixed hardwood/conifer forests at lower elevations to mixed conifer and subalpine forests at mid- to upper-elevations. Some portions of this section of the PCT, particularly further to the south and to the east, consist of semiarid sagebrush (*Artemisia tridentata*) and montane chaparral (Schoenherr, 1992).

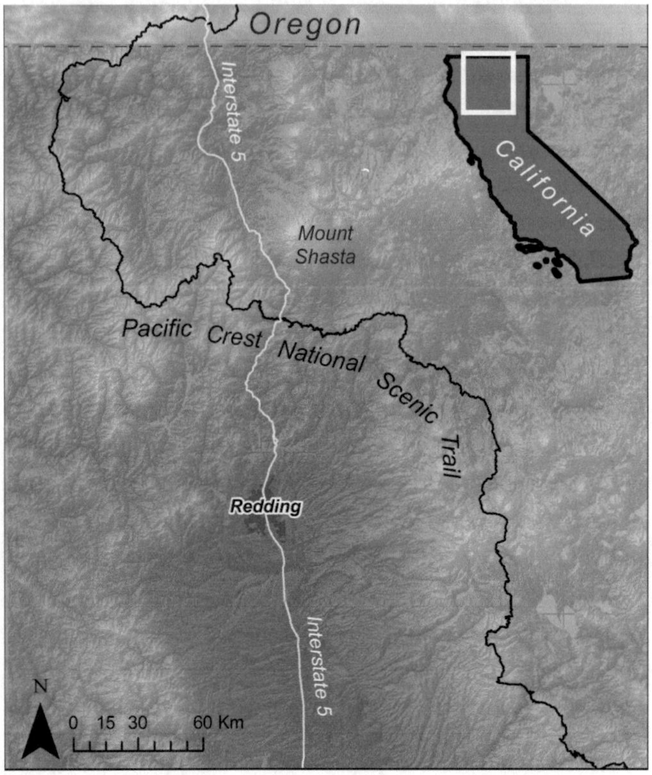

Figure 1. The PCT route where data was collected in 2010 by McGrann and Furnas (2016).

Occurrence Data

The PCT Data contains avian occurrence data for the year 2010 and was acquired from McGrann and Furnas (2016). The PCT Data was collected along the trail via fixed-radius (50 m) point-counts and automated recorders in a standardized method as described in detail in Furnas & Callas (2015), McGrann and Furnas (2016), and Furnas and McGrann (2018). BKHGRO was detected at 83 sites, YERWAR, 206 sites, and HAIWOO, 29 sites.

We downloaded data from GBIF and BISON for the year 2010 in the months of May, June, and July to match the timeframe of when surveys were completed for the PCT Data. These two databases both obtain occurrence data from data contributors such as Cornell lab of Ornithology, the eBird Observation Dataset, and the Great Backyard Bird Count. These two datasets were combined into a single dataset, which we call GBIF-BISON Data. For each species, the number of individual detections extracted from the GBIF-BISON Data for the study area included 84 for BKHGRO, 195 for YERWAR, and 101 for HAIWOO. The GBIF-BISON Data contains spatial bias due to an uneven method of sampling and may distort spatial models (Beck et al., 2014). Modeling was completed using the GBIF-BISON Data and the PCT Data separately and combined to cover a larger area for each species in the study area, which we call the Combined Dataset.

Environmental Covariates

We tested 14 environmental covariates that we thought would be associated with habitat for the three avian species based upon the natural history and habitat requirements as described in Sousa (1987) and DeGraaf & Rappole (1995). We downloaded 8 variables from the Oregon State University's Landscape Ecology, Modelling, Mapping and Analysis (LEMMA) program, which obtain their data by integrating vegetation measurements from field surveys, mapped environmental data, and Landsat Thematic Mapper (TM) imagery (30 m resolution) (Ohmann & Gregory, 2002; Landscape Ecology, Modeling, Mapping, 2020). The variables we selected were hardwood and conifer canopy cover, total canopy cover, quadratic mean diameter of all dominant and codominant trees (qmd_dom), forest type based on the basal area of dominant tree species (Fortypba), vegetation class based on the canopy cover and basal area (vegetation class), and density of live trees and snags. The Fortypba layer contained 983 categorical values, but some modeling software, such as the Hyper-Envelope Modeling Interface Version 2 (HEMI2), require less than 255 categories, so we did the following process to reduce the amount into coarser scale. First, we extracted the Fortypba values to the survey sites and classified these as values from 1-98. The remaining values in the Fortypba layer were combined into coarser classifications based on the dominant tree species. At the end of this process, 141 categories were represented in the new Fortypba layer.

The remaining 6 covariates included in our models were distance from fire, distance from water, distance from bark beetle infestation, elevation (as surrogate for temperature), aspect, and slope. Distance from fire, water, and bark beetle infestations help measure habitat resources (e.g., food, cover, and nesting habitat) that may be important to a species. Elevation, aspect, and slope are characteristics of the terrain which can be associated with temperature or light exposure, which also influences vegetation. All covariates were converted to rasters of grid cells that represent the value of the covariate at on location in the landscape. We calculated a distance to fire raster (i.e., a grid of cells where the value assigned to each cell represents a distance value) by downloading fire polygons of mapped burned areas from the Monitoring Trends in Burn Severity data set (Eidenshink et al., 2009) and applying the Euclidian distance function in ArcGIS (version 10.7.1, Environmental Systems Research Institute, Redlands, CA). This resulted in a raster that calculated distance to the edge of the polygon where everything within the polygon was assigned a value of zero. Stream and river polylines were downloaded from the National Hydrography Dataset (1:24,000; Terziotti & Archuleta, 2020). Bark beetle infestation polygons were downloaded from the USDA Forest Service and were obtained via aerial "sketchmapping" (Schrader-Patton & Pywell, 2003). We used a similar approach to convert streams and rivers polylines into a distance to water raster and to convert bark beetle infestation polygons into a distance raster. Digital elevation maps (DEMs) were downloaded from the U.S. Geologic Survey (30 m resolution). Temperature decreases with increasing elevation according to a known rate (i.e., the adiabatic lapse rate). Therefore, we considered elevation to be a surrogate covariate for temperature. From the DEMs, we also derived aspect and slope rasters, which represent a surrogate covariate for precipitation (Geroy et al., 2011; Phillips & Schümm, 1987). All rasters were scaled to 30 m to match the vegetation covariate raster cell size, clipped to the study area, and converted into ASCII files using ArcGIS.

We reduced the number of variables used to create the model by performing several steps. First, we analyzed the correlation between all the environmental variables using the Pearson correlation statistic (Appendix: R Script). Next, we used MaxEnt's jackknife feature to evaluate each environmental variables contribution to each model (Elith et al., 2010). We removed variables that had less than 2% contribution and were highly correlated (i.e., a correlation coefficient > 0.7), or did not impact the jackknife's regularization training gain when removed. The regularization training gain is a measure between a random sample of the entire study area the species could inhabit and the environmental covariates correlated to the species occurrence (Elith et al., 2010).

MaxEnt

We performed MaxEnt within the software BlueSpray (beta version 42, SchoonerTurtles, Arcata, CA), which calculates area under the curve (AUC) and the Akaike information criterion (AIC). AUC measured model performance by measuring a model's discriminatory ability and represents the proportion of times the actual sample of presence locations has a larger estimated suitability than a random sample (Fielding & Bell, 1997). AIC attempts to balance predictive ability of the model with model complexity by providing an estimate of the relative "quality" among a series of competing models (Plant, 2012). Additionally, BlueSpray can perform Monte Carlo simulations within MaxEnt (Graham & Kimble, 2018). Monte Carlo simulations are a statistical method where the model is replicated a large amount of times with aspects of the model randomized with each replicate (Plant, 2012). Using the spatial coordinates of the bird occurrences and the set of covariates that we selected, which were selected based upon the criterion of at least 2% contribution to the model, we increased the regularization multiplier in the combined model by increments of 0.5 until we achieved the lowest AIC. A higher regularization multiplier smooths out the response curves to reduce the complexity of the models produced. To create the most parsimonious model and to be able to evaluate how the two datasets models compare for each species, we used the best regularization parameter from the Combined Dataset to create models with only the PCT Data and only the GBIF-BISON Data.

Model Selection and Evaluation

Models were evaluated based on their AIC (Muscarella et al., 2014), delta Akaike information criteria (ΔAIC), and AUC. In order assess whether the models are accurately predicting suitable habitat, we calculated the number of observed occurrences that fell within the predicted habitat suitability grid cells using the 10% logistic threshold MaxEnt calculated for each species. This 10% logistic threshold indicates probability value that is the minimum value for suitable habitat and it can assist in determining the generality of our models between datasets.

We also employed cross-validation to test the generality of our models across datasets and to evaluate model performance on the best model for each species with the lowest AIC. Cross-validation can be performed in MaxEnt. This process involves splitting a designated percent of occurrence locations into a training dataset, which is used to fit the model, and a testing dataset, which is used to test against the rest of the occurrence locations (Merow et al., 2013). A robust model would have little variation of predicted habitat among iterations (Kimble, 2016) and generally, models that over fit the data perform well on the training data and poorly on the test data. For the Combined Dataset, 70% of the data was used for training and 30% was used for testing. We performed cross-validation on the PCT Data where 100% of the data was used for training and used the GBIF-BISON Data for testing. We then reversed the process using 100% of the GBIF-BISON for training and used the PCT Data for testing. To assess whether the models are accurately predicting suitable habitat, we calculated how many of these occurrences fall within the 10% logistic threshold.

To further validate model robustness, we used Monte Carlo simulations to check for spatial uncertainty in the occurrence points and covariates. We injected error into each of the species models with the Combined Datasets using the Monte Carlo feature in BlueSpray (Graham & Kimble, 2018). We ran 80 iterations and evaluated the mean AIC, standard deviation of the AIC, mean AUC and standard deviation of the AUC. The DEM is noted to vary vertically up to 2.42 meters in the conterminous United States (Gesch et al., 2014). In the programming language Python, we calculated the standard deviation of error in the slope and aspect rasters by varying the amount of error in the DEM and taking the average standard deviation over 10 runs (Appendix: Python Script). We found that slope had an average standard error of 0.728 degrees and aspect varied by 56.91 degrees. Data from OSU LEMMA underestimates values (Bell et al., 2015), with most rasters seen to have reductions by 0.05. Bark beetle infestations had patch areas combined into a larger polygon (USDA, 2010), so there may be an overestimation. LEMMA also notes that their overall classification accuracy for 10 categories was 45% and that most misclassification errors were minor (Ohmann & Gregory, 2002). Monitoring Trends in Burn Severity map fires accurately that are greater than 1000 acres (Eidenshink et al., 2009). GPS average error is 0.715 meters (U.S. Department Of Defense, 2007).

Results

Six environmental covariates for BKHGRO, six for HAIWOO, and five for YERWAR contributed to explaining more than 2% of the variation in the MaxEnt model initially made for each species and were included in building these individual models for each species using the Combined Dataset (Table 1). These covariates appear to have a significant correlation with the occurrence locations of the species

Table 1. Environmental covariates selected for each species to build the habitat suitability models based on the MaxEnt Jackknife feature using the Combined Dataset.

Species	Environmental Covariates
BKHGRO	1) Fortypba 2) elevation 3) distance to bark beetle infestations 4) distance to fires 5) slope 6) distance to streams
HAIWOO	1) Fortypba 2) distance to bark beetle infestations 3) slope 4) hardwood canopy cover 5) distance to fires 6) aspect
YERWAR	1) Fortypba 2) slope 3) hardwood canopy cover 4) distance to fires 5) distance to streams

and were determined to be valuable in creating the habitat suitability models.

The best model for BKHGRO, based on the lowest AIC, had a regularization multiplier of 1.5 to produce smooth response curves that did not over fit the data. For HAIWOO, the regularization multiplier was 5.5. For YERWAR, the best regularization multiplier was 1. In the best model of the Combined Dataset for each species, the 10% logistic threshold MaxEnt calculated for each species was 0.33 for BKHGRO, 0.37 for HAIWOO, and 0.28 for YERWAR (Table 2).

The PCT Data and the GBIF-BISON Data had consistently higher AUC values than the Combined Dataset (Table 3). The Combined Dataset was lower in AUC by 2-3 units. When we performed cross-validation on the PCT Data or the GBIF-BISON Data, the AUC decreased to just over the random prediction line. When we performed cross-validation on the Combined Dataset, the AUC decreased slightly.

Table 2. Model parameters run for each bird species and their resulting AIC, ΔAIC, and AUC.

Data	Bird Species	Software	Regularization multiplier	AIC	ΔAIC	AUC
Combined	BKHGRO	MaxEnt	1.5	10157	0	0.88
Combined	BKHGRO	MaxEnt	2	10159	2	0.87
Combined	BKHGRO	MaxEnt	1	10201	44	0.89
Combined	HAIWOO	MaxEnt	5.5	4782	0	0.75
Combined	HAIWOO	MaxEnt	6	4784	2	0.75
Combined	HAIWOO	MaxEnt	5	4785	3	0.76
Combined	HAIWOO	MaxEnt	4.5	4789	7	0.76
Combined	HAIWOO	MaxEnt	4	4796	14	0.76
Combined	HAIWOO	MaxEnt	3.5	4800	18	0.77
Combined	HAIWOO	MaxEnt	3	4825	43	0.78
Combined	HAIWOO	MaxEnt	2.5	4896	114	0.79
Combined	HAIWOO	MaxEnt	2	4968	186	0.80
Combined	HAIWOO	MaxEnt	1.5	4986	204	0.82
Combined	HAIWOO	MaxEnt	1	5110	328	0.83
Combined	YERWAR	MaxEnt	1	29863	0	0.80
Combined	YERWAR	MaxEnt	1.5	29866	3	0.79

BKHGRO Habitat Suitability

We found that covariates elevation, Fortypba, distance to bark beetle infestations, slope, stream distance, and fires contributed significantly to our model for BHKGRO habitat suitability (Figs. 2 & 3). BKHGRO appears to find close proximity to bark beetle infestations as more suitable; this species has been observed to prefer forests impacted by beetle impacts (Fair et al., 2018; Mosher, 2011). BKHGRO most likely does not forage on the boring beetles but may instead feed on other insects located in areas impacted by bark beetles, since BKHGRO glean insects (Airola & Barrett, 1985) rather than drill for boring beetles. The species appears to favor elevations above 800 meters but below 1,100 meters, but our models indicate

Table 3. Results of models created with the two datasets and the Combined Dataset. Cross-validation was performed against each of the models along with their AUC. For the logistic threshold, the number of occurrences that fall within the area selected by the model divided by the number of occurrences available by the dataset is provided.

Bird Species	Data Used to Create Model	AUC	Data used for Cross-validation	Cross-validation AUC	Logistic Threshold for the PCT Data	Logistic Threshold for the GBIF-BISON Data
BKHGRO	PCT	0.95	GBIF- BISON	0.55	0.88	0.12
BKHGRO	PCT	0.91	PCT	0.85	0.85	0.30
BKHGRO	GBIF-BISON	0.91	PCT	0.56	0.23	0.88
BKHGRO	GBIF-BISON	0.90	GBIF- BISON	0.78	0.28	0.87
BKHGRO	Combined	0.88	Combined	0.83	0.93	0.88
HAIWOO	PCT	0.78	GBIF- BISON	0.57	1.00	0.83
HAIWOO	PCT	0.77	PCT	0.55	0.97	0.88
HAIWOO	GBIF/BISON	0.79	PCT	0.58	0.55	0.83
HAIWOO	GBIF-BISON	0.79	GBIF- BISON	0.77	0.59	0.89
HAIWOO	Combined	0.76	Combined	0.71	0.83	0.91
YERWAR	PCT	0.91	GBIF- BISON	0.51	0.85	0.24
YERWAR	PCT	0.91	PCT	0.88	0.83	0.25
YERWAR	GBIF-BISON	0.81	PCT	0.54	0.56	0.92
YERWAR	GBIF-BISON	0.82	GBIF- BISON	0.72	0.52	0.89
YERWAR	Combined	0.79	Combined	0.77	0.94	0.87

some suitability in lower elevations where occurrences were recorded in isolated forested habitats in the central valley as indicated by our Fortypba layer. We found higher suitability closer to burned areas. Suitability also peaks close to streams. BKHGRO has been noted to have a preference for a mixed hardwood/conifer plant community (Airola & Barrett, 1985) and Fortypba did confirm these preferences showing a high affinity for white fir (*Abies concolor*), Ponderosa pine (*Pinus ponderosa*), California incense cedar (*Calocedrus decurrens*), sugar pine (*Pinus lambertiana*), Douglas fir (*Pseudotsuga menziesii*), California black oak (*Quercus kelloggii*), and canyon live oak (*Quercus chrysolepis*). This species appears to find slopes less than 30 degrees as more suitable.

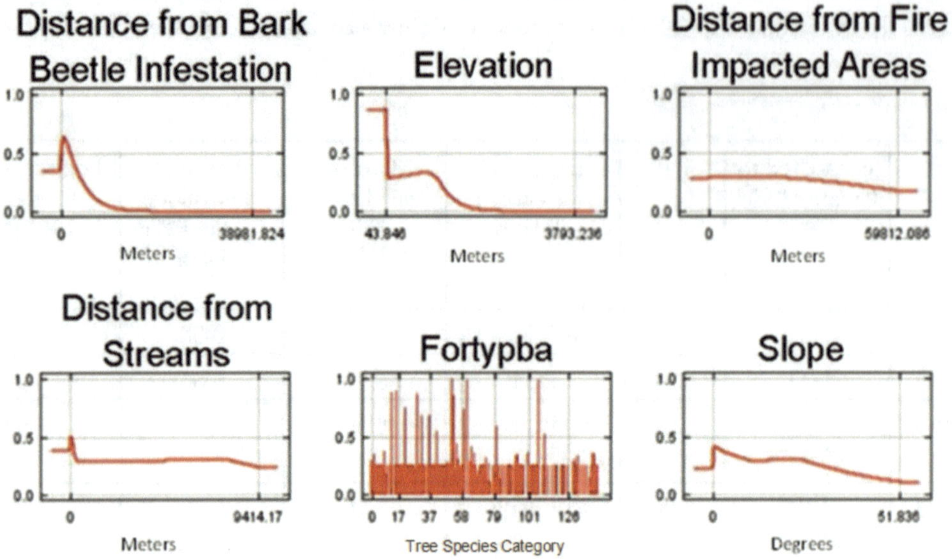

Figure 2. BKHGRO response curves from MaxEnt with a regularization multiplier of 1.5 for each of the environmental covariates using the Combined Dataset.

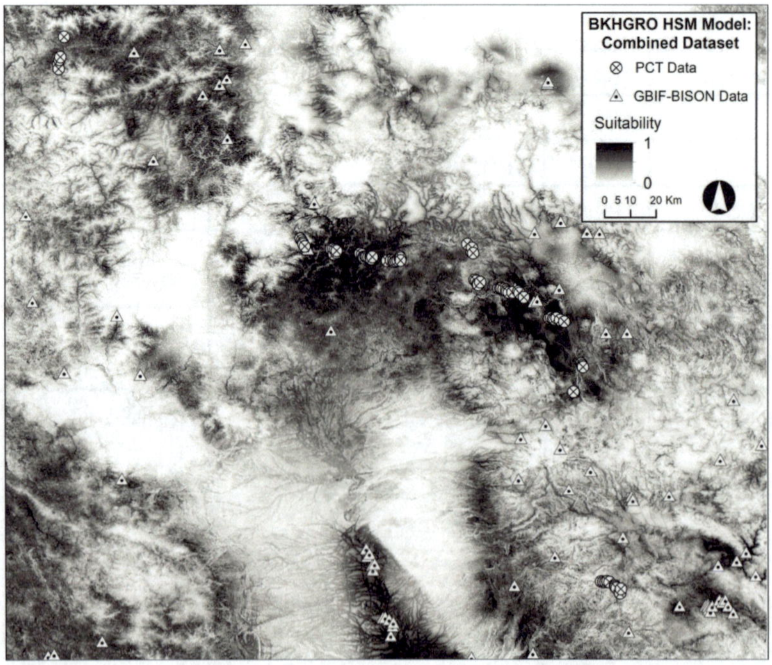

Figure 3. Habitat suitability model for the BKHGRO built from the Combined Dataset with a regularization multiplier of 1.5.

HAIWOO Habitat Suitability

For HAIWOO, we found that the covariates of aspect, distance from bark beetle infestations, distance from fires, hardwood canopy cover, Fortypba, and slope contributed significantly to our model of habitat suitability (Figs. 4 & 5), where areas closer to bark beetle infestations and fire-impacted habitat are considered more suitable. HAIWOO appears to prefer northwest facing slopes that are under 30 degrees. Areas with lower percentages of hardwood canopy, but greater percentages of conifer species, had higher suitability, including red fir (*Abies procera*), white fir, Jeffrey pine (*Pinus jeffreyi*), Ponderosa pine, white oak (*Quercus garryana*), and California black oak.

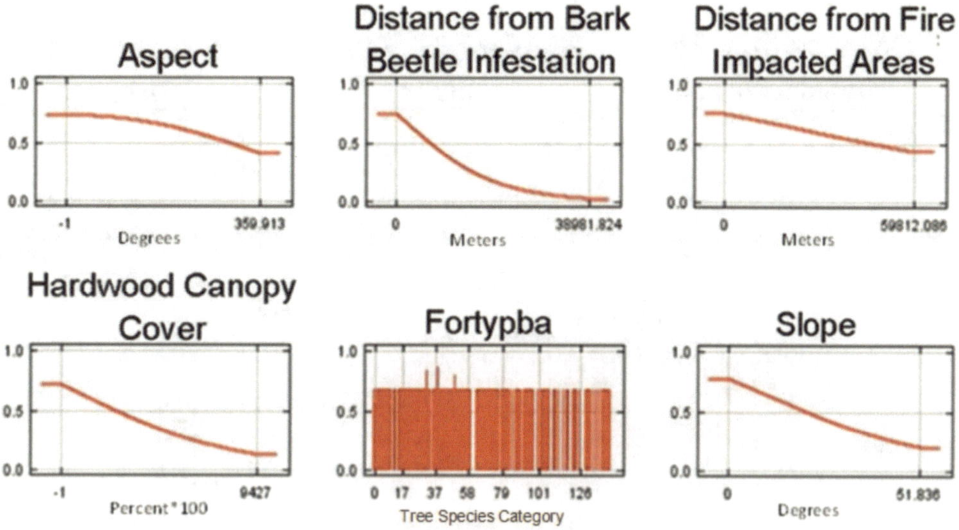

Figure 4. HAIWOO response curves from MaxEnt with a regularization multiplier of 5.5 for each of the environmental covariates using the Combined Dataset.

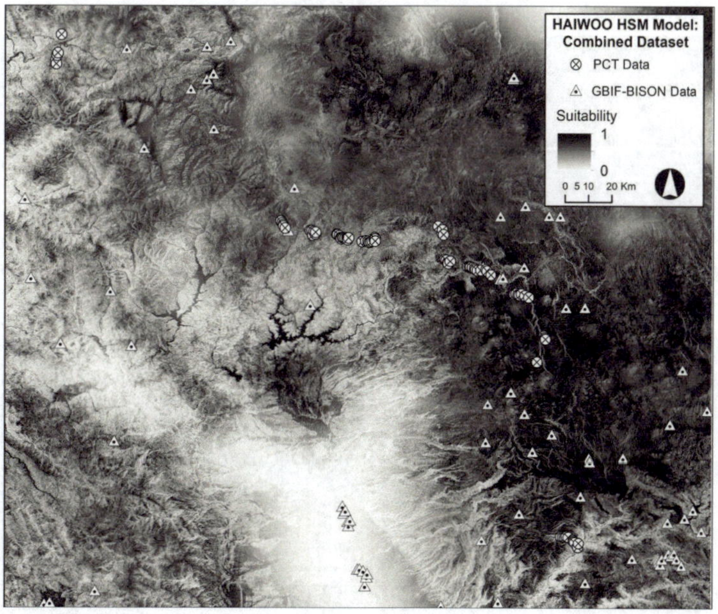

Figure 5. Habitat suitability model for the HAIWOO built from the Combined Dataset with a regularization multiplier of 5.5.

YERWAR Habitat Suitability

For YERWAR, we found that the covariates of distance from fire impacted areas, hardwood canopy cover, distance from streams, Fortypba, and slope contributed significantly to our model of habitat suitability (Fig. 6 & 7). Our model also indicated higher suitability nearer to areas impacted by fire. We found that suability was greatest with little to no hardwood canopy cover. Areas nearer to streams have greater suitability than areas away from streams. Suitability was greatest in habitats dominated by coniferous trees, including white fir, red fir, California incense cedar, Jeffrey pine, sugar pine, Western white pine (*Pinus monticola*),

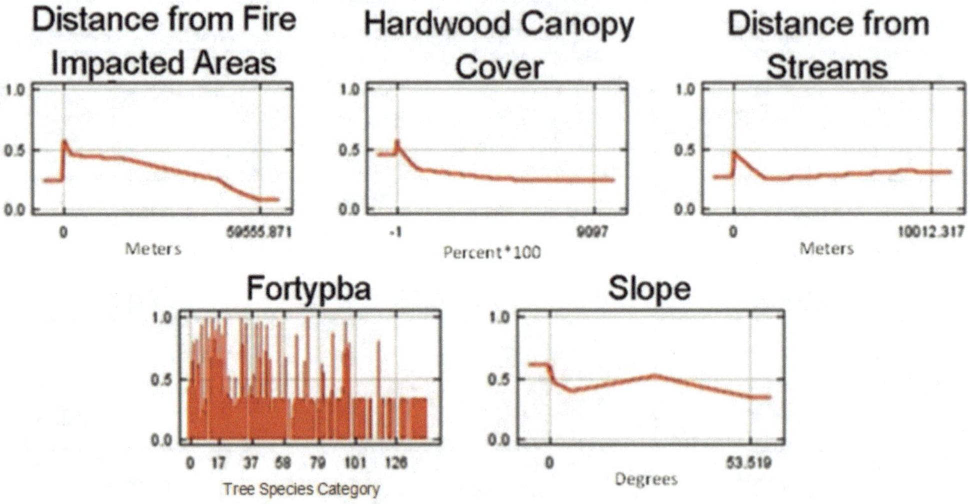

Figure 6. YERWAR response curves from MaxEnt with a regularization multiplier of 1 for each of the environmental covariates using the Combined Dataset.

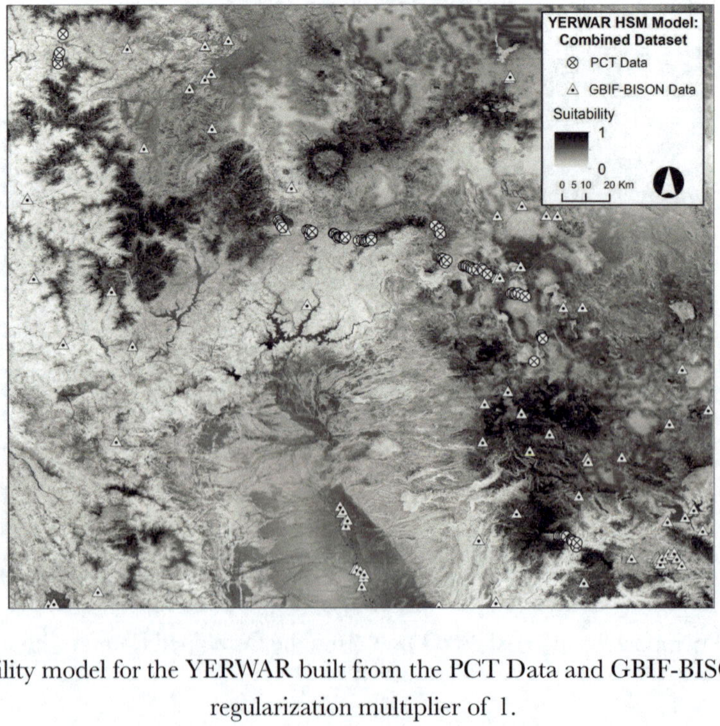

Figure 7. Habitat suitability model for the YERWAR built from the PCT Data and GBIF-BISON Data combined with a regularization multiplier of 1.

Ponderosa pine, Douglas fir, mountain hemlock (*Tsuga mertensiana*), knobcone pine (*Pinus attenuate*), Brewer spruce (*Picea breweriana*), lodgepole pine (*Pinus contorta*), live oak, subalpine fir (*Abies lasiocarpa*), white oak, bitter cherry (*Prunus emarginata*), blue oak (*Quercus douglasii*), and Pacific silver fir (*Abies amabilis*). YERWAR appears to find slopes less than 30 degrees as more suitable.

Model Evaluation

With injected uncertainty into the best MaxEnt models for BKHGRO, the AIC increased by 58 and the AUC decreased by 0.03 from the original model. For HAIWOO, the AIC increased by 19 and the AUC decreased by 0.04. For YERWAR model, the AIC increased by 135 and the AUC decreased by 0.02.

Discussion

Our approach of combining complementary datasets derived from different methods in the field has resulted in habitat suitability models that are generalizable across our entire study area. Previous studies have also combined this PCT Data with data from another large-scale biodiversity monitoring program throughout northern California to model bird diversity-climate relationships (McGrann & Furnas, 2016; Furnas & McGrann 2018). In these studies, the PCT Data served to strengthen the multi-species occupancy models used for the analysis and to improve the representation of the remote, high-elevation habitats that were poorly sampled by the more extensive region-wide monitoring program. The models created with the Combined Dataset performed at predicting the occurrences in all but one case with HAIWOO. In a similar approach, we used MaxEnt to create response curves that represent the expected habitat associations for each species across the entire study area beyond only the PCT. Further, we determined that our models were robust after injecting error into the observed data and the covariates and found the AUC only decreased by 0.02 to 0.04 points. Altogether, this indicates that our models are predictive of the actual spatial distribution of these species and of where these species might find suitable habitat across our entire study area. Our future research direction will include the use of these models to study how these species' habitats might be shifting due to drought, fire, or climate change, thus making our modeling approach useful to management decisions.

Table 3. Monte Carlo results for injecting uncertainty into the model for BKHGRO.

Species		**AIC Deviation**		**AUC Deviation**
BKHGRO	10215	63.99	0.85	0.01
HAIWOO	4801	28.45	0.71	0.01
YERWAR	29998	84.20	0.78	0.01

By combining the two datasets, we created a model that increased the ability to predict locations that included occurrences with a small reduction in AUC values. Although the AUC for the PCT Data was high, it did not predict occurrences well beyond the trail for YERWAR and BKHGRO, indicating that the PCT Data may be suitable for fine-scale analysis on species distributions along this specific region of the trail. The PCT survey sites, due to the trail's design and the route chosen for the trail, may be biased towards higher elevations and other habitats disproportionately occurring along the trail, which was also reported by Furnas & McGrann (2018). It appears, from our analysis, that the PCT Data does not have enough predictive power to project to a larger area and requires additional data covering the broader range of environmental covariates throughout the study area. Yet, it would be cost prohibitive to apply more widely, across the entire study area, the intensive survey methods designed for application along this transect.

The GBIF-BISON Data has wider coverage of the study area but also has its own inherent biases. The AUC for the GBIF-BISON Data was slightly lower than the PCT Data overall, and it did not project well to the PCT Data. This can indicate that although GBIF-BISON Data may cover a larger area, it may not have the consistent and structured sampling design required for fine scale modeling. Additionally, data from these sources are generally biased towards roads (Ronen Kadmon, 2004) and contain surveyor bias as observers favor habitats that are easier to access from roads or that are considered favorable for observing greater numbers of birds (Tulloch & Szabo, 2012). For example, we noticed that HAIWOO, which was expected to benefit from foraging in fire-impacted areas, was observed to have few occurrences within these areas (Figure 8). We suspect that observers who collected GBIF-BISON Data, chose to avoid burned forests. The GBIF-BISON Data, however, included some sampled areas at lower elevations, such as in the Sacramento Valley, where the PCT Data did not have any survey sites. The GBIF-BISON Data also had gaps, particularly in the remote and high-elevation wilderness areas, where the PCT Data was able fill in. Inclusion of the occurrences in the Sacramento Valley influenced the shape of the response curves, and the full range of the species was represented more appropriately.

Habitat Associations for BKHGRO, HAIWOO, and YERWAR

Response curves described well our anticipated habitat associations for BKHGRO, HAIWOO, and YERWAR. Previous studies have documented BKHGRO in disturbed habitat near fires (Bagne & Purcell, 2011) and, more rarely, in high-elevation habitats (Wilson, 2013). As a canopy nester, it prefers nesting in close proximity to streams, which can act to moderate temperatures for the nest site (Becker, 2013). BKHGRO forage in a variety of habitats (Airola & Barrett, 1985) but within 2 km of a water source and readily use shrubs in early successional habitats (Gardali & Holmes, 2011). BKHGRO's affinity for shrubs may also lead to an affinity for a specific tree cover from the Fortypba layer (Pase, 1982), particularly at lower- to mid-elevation montane forests where a distinct shrub layer is commonplace in the understory. The response curves for the BKHGRO, when run with individual covariates, may suggest a bimodal response with distance to bark beetle infestations (Figure 8). They are found in bark beetle infested habitats (Mosher, 2011) where they may consume arthropods that follow a bark beetle infestation (Weslien &

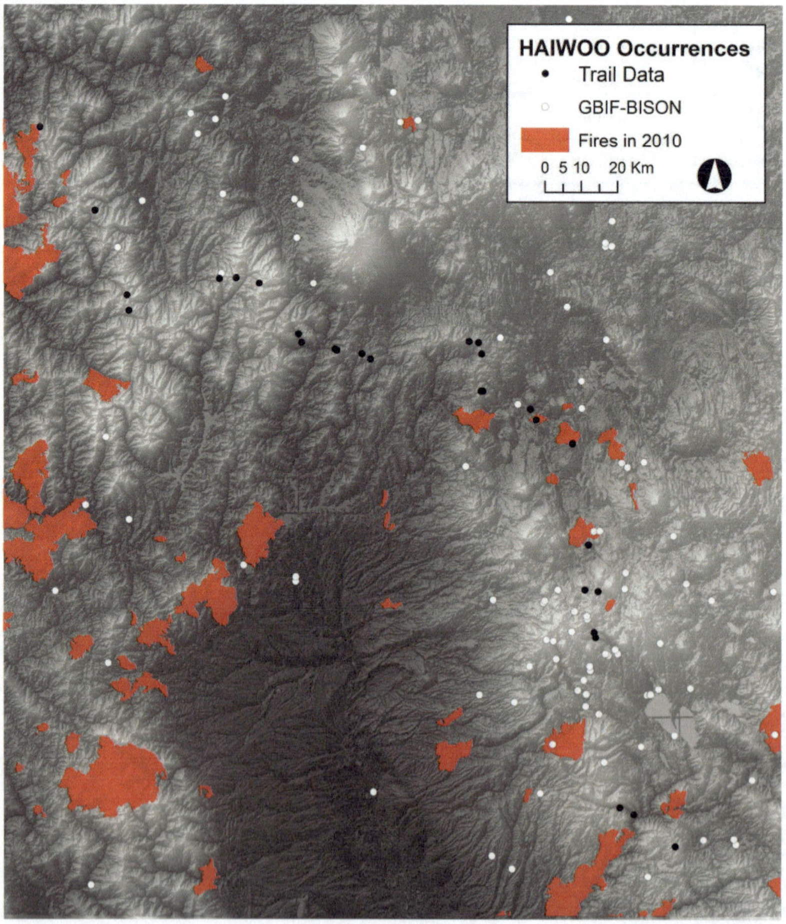

Figure 8. Occurrences from HAIWOO and the zones of fire-impacted areas. Fire polygons from Monitoring Trends in Burn Severity (2009).

Schroeder, 1999), but also forage away from these infestations on other insects or seeds. When the distance to bark beetle infestations covariate is combined into the full model, it takes on a curve where close proximity to bark beetles is very suitable.

HAIWOO utilizes fire-impacted areas and forages on bark beetles (Saab et al., 2019), and we observed these habitats to have higher suitability. Since HAIWOO nest and forage in both snags or live conifer trees that may show signs of defoliation from bark beetles (Bull et al., 1986), lower amounts of hardwood canopy cover would have higher suitability. Slope and aspect may influence the woodpecker's choice of nest site, preferring cooler, moister areas (Bull et al., 1986). All the categories for tree species, from the Fortypba layer, showed some suitability, with only three tree species showing higher suitability. This could indicate that HAIWOO is a generalist, preferring many tree species for foraging and nesting.

For YERWAR, high suitability was evident near, but not *within*, fire-impacted habitat, as the edge of this habitat may provide good forage for insects that are abundant on the edges of burned forests. We documented higher suitability in close proximity to streams; which is also documented by others (Kirkpatrick et al., 2009).

Conclusions

We found that first identifying the trade-offs of each dataset, and deciding on whether to select the appropriate dataset, or combine both, was crucial in creating the best model to address our research questions and hypotheses. The PCT Data, derived from a transect survey protocol, may be suitable for intensive sampling along environmental gradients and in describing climate-diversity relationship (the original purpose of the data) but not necessarily for extrapolation far beyond the trail itself. The intent of the original biodiversity study along the PCT was *not* to describe species distributions across a broader study area. Data from BISON or GBIF is beneficial for broad scale analysis but may leave out areas for fine scale analysis and these data may poorly represent more remote and high-elevation habitats. Also, the GBIF-BISON Data may not include the full range of the species and their habitats. If we use these models to generate fine-scale maps, we need to find additional datasets that capture more of the covariates required to model those areas, and combining datasets, habitat suitability models can be built that fill in spatial gaps and can more adequately inform management (Turner et al., 2016). Additionally, we found that BISON-GBIF underrepresented fire-impacted areas and throughout the region.

We provide a useful approach to modeling habitat suitability by combining an intensive transect survey with data sources from repositories such as BISON and GBIF. Overall, our models well characterized the suitability of habitat for these three bird species and demonstrated their utility. With well-developed models, managers can determine which habitats in their area are suitable and also track the overall response of avian diversity to drought, fire, climate change, or human land-use change. Future research directions should examine whether this modeling approach can track the response of birds, and their habitats, to these environmental changes.

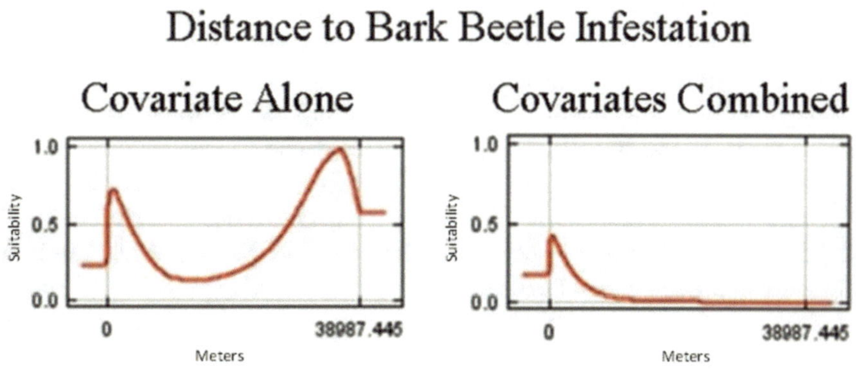

Figure 9. Response curves for distance to bark beetle infestations with the covariate alone and combined in the full model. The covariate alone displays a bimodal response where the species is either very close to the infested area, or is further away. The covariate combined shows that areas near the infested habitat is more suitable.

Acknowledgements

General Acknowledgements: We would like to thank Matthew Gregory at OSU LEMMA for his work with assisting in covariate data collection. We would also like to thank Matthew Johnson at Humboldt State University for his consultation at the data analysis stage. Data Contributors: We thank the numerous data contributors whose public data made this project possible. Oregon State University contributed large rasters on forest structure from their LEMMA program to aide in this research. Multiple agencies contribute to MTBS data on fire severity and area coverage. United States Geologic Survey allowed open access to detailed DEMs. BISON and GBIF contributed avian occurrence data. Author Contributions: H.P. designed methods, collected data, and conducted research; M.M. provided avian data from the PCT and substantial funding for avian recording interpretations; J.G. supervised research; H.P. wrote the paper; and M.M. and J.G. provided comments and revised the paper. Data Repository: Analyses reported in this article can be reproduced using the data provided by McGrann and Furnas (2016) and is stored at the PCT Biodiversity Megatransect data repository.

Literature Cited

Airola, D. A., & Barrett, R. H. (1985). Foraging and Habitat Relationships of Insect-Gleaning Birds in a Sierra Nevada Mixed- Conifer Forest. *The Condor*, *87*(2), 205–216.

Asner, G. P., Brodrick, P. G., Anderson, C. B., Vaughn, N., Knapp, D. E., & Martin, R. E. (2015). Progressive forest canopy water loss during the 2012-2015 California drought. *Proceedings of the National Academy of Sciences of the United States of America*, *113*(2), E249–E255. https://doi.org/10.1073/pnas.1523397113

Austin, M. P., & Heyligers, P. C. (1989). Vegetation survey design for conservation: Gradsect sampling of forests in North-eastern New South Wales. *Biological Conservation*, *50*(1–4), 13–32. https://doi.org/10.1016/0006-3207(89)90003-7

Bagne, K. E., & Purcell, K. L. (2011). Short-term responses of birds to prescribed fire in fire-suppressed forests of California. *Journal of Wildlife Management*, *75*(5), 1051–1060. https://doi.org/10.1002/jwmg.128

Barry, S., & Elith, J. (2006). Error and uncertainty in habitat models. *Journal of Applied Ecology*, *43*(3), 413–423. https://doi.org/10.1111/j.1365-2664.2006.01136.x

Beck, J., Böller, M., Erhardt, A., & Schwanghart, W. (2014). Spatial bias in the GBIF database and its effect on modeling species' geographic distributions. *Ecological Informatics*, *19*, 10–15. https://doi.org/10.1016/j.ecoinf.2013.11.002

Becker, M. (2013). *The Role of Urban Climate and Land Cover in Phenology, Nest Success, and Habitat Use* (Issue August). University of Nevada, Reno.

Bell, D. M., Gregory, M. J., Roberts, H. M., Davis, R. J., & Ohmann, J. L. (2015). How sampling and scale limit accuracy assessment of vegetation maps: A comment on Loehle et al. (2015). *Forest Ecology and Management*, *358*, 361–364. https://doi.org/10.1016/j.foreco.2015.07.017

Bull, E. L., Peterson, S. R., & Thomas, J. W. (1986). *Resource partitioning among woodpeckers in northeastern Oregon.*

DeGraaf, R. M., & Rappole, J. H. (1995). *Neotropical migratory birds: natural history, distribution, and population change*. Comstock Publishing Associates.

Eidenshink, J., Schwind, B., Brewer, K., Zhu, Z.-L., Quayle, B., & Howard, S. (2009). A Project for Monitoring Trends in Burn Severity. *Fire Ecology*, *3*(1), 3–21. https://doi.org/10.4996/fireecology.0301003

Elith, J., Phillips, S. J., Hastie, T., & Dudı, M. (2010). A statistical explanation of MaxEnt for ecologists. *Diversity and Distributions*, 1–15. https://doi.org/10.1111/j.1472-4642.2010.00725.x

Fair, J. M., Hathcock, C. D., & Bartlow, A. W. (2018). Avian communities are decreasing with piñon pine mortality in the southwest. *Biological Conservation*, *226*(June), 186–195. https://doi.org/10.1016/j.biocon.2018.06.035

Fielding, A. H., & Bell, J. F. (1997). A review of methods for the assessment of prediction errors in conservation presence/absence models. *Environmental Conservation*, *24*(1), 38–49. https://doi.org/10.1017/S0376892997000088

Furnas, B. J., & Callas, R. L. (2015). Using automated recorders and occupancy models to monitor common forest birds across a large geographic region. *Journal of Wildlife Management*, *79*(2), 325–337. https://doi.org/10.1002/jwmg.821

Furnas, B. J., & McGrann, M. C. (2018). Using occupancy modeling to monitor dates of peak vocal activity for passerines in California. *The Condor*, *120*(1), 188–200. https://doi.org/10.1650/condor-17-165.1

Gardali, T., & Holmes, A. L. (2011). Maximizing benefits from riparian revegetation efforts: Local- and landscape-level determinants of avian response. *Environmental Management*, *48*(1), 28–37. https://doi.org/10.1007/s00267-011-9623-6

Geroy, I. J., Gribb, M. M., Marshall, H. P., Chandler, D. G., Benner, S. G., & McNamara, J. P. (2011). Aspect influences on soil water retention and storage. *Hydrological Processes*, *25*(November), 3836–3842. https://doi.org/10.1002/hyp.8281

Gesch, D. B., Oimoen, M. J., & Evans, G. A. (2014). Accuracy Assessment of the U.S. Geological Survey National Elevation Dataset, and Comparison with Other Large-Area Elevation Datasets-SRTM and ASTER Open-File Report 2014-1008. In *U.S. Geological Survey Open-File Report 2014-1008*. https://doi.org/10.3133/ofr20141008

Graham, J., & Kimble, M. (2018). Visualizing uncertainty in habitat suitability models with the hyper-envelope modeling interface, version 2. *Ecology and Evolution, December 2017*, 251–264. https://doi.org/10.1002/ece3.4720

Hedley, S. L., & Buckland, S. T. (2004). Spatial models for line transect sampling. *Journal of Agricultural, Biological, and Environmental Statistics*, *9*(2), 181–199. https://doi.org/10.1198/1085711043578

Kadmon, R., Farber, O., & Danin, A. (2016). A Systematic Analysis of Factors Affecting the Performance of Climatic Envelope Models Author (s): Ronen Kadmon , Oren Farber and Avinoam Danin Published by : Ecological Society of America Stable URL : http://www.jstor.org/stable/4134701 REFERENCES Link. *Ecological Applications*, *13*(3), 853–867.

Kimble, M. S. (2016). *MAPPING UNCERTAINTY OF HABITAT SUITABILITY MODELS OF FOUR NORTH AMERICAN TREE SPECIES*. Humboldt State University.

Kirkpatrick, C., Conway, C. J., & LaRoche, D. (2009). *Surface Water Depletion and Riparian Birds* (Issue June).

Klicka, J., Spellman, G. M., Winker, K., Chua, V., & Smith, B. T. (2011). A Phylogeographic and population genetic analysis of a widespread, sedentary North American bird: The Hairy Woodpecker (Picoides villosus). *Auk*, *128*(2), 346–362. https://doi.org/10.1525/auk.2011.10264

Landscape Ecology Modeling, Mapping, and A. (LEMMA) T. (2020). *Gradient Nearest Neighbor (GNN) raster dataset (version 2020.01). Modeled forest vegetation data using direct gradient analysis and nearest neighbor imputation.* https://lemmadownload.forestry.oregonstate.edu

Lee, A. P., Ding, T., Hsu, F., Geng, S., Lee, P., Ding, T., Hsu, F., & Geng, S. (2004). Breeding Bird Species Richness in Taiwan : Distribution on Gradients of Elevation , Primary Productivity and Urbanization Published by : Wiley Stable URL : https://www.jstor.org/stable/3554659 - Breeding bird species richness in Taiwan : W W5 distribution. *Journal of Biogeography*, *31*(2), 307–314.

MacArthur, R. H. (1965). Patterns of species diversity. *Biological Reviews*, *40*(4), 510–533.

McGrann, M. C., & Furnas, B. J. (2016). Divergent species richness and vocal behavior in avian migratory guilds along an elevational gradient. *ECOSPHERE*, *7*(August), 1–14.

McGrann, M. C., & Thorne, J. H. (2014). Elevation Ranges of Birds along California's Pacific Crest Trail. *Western Birds*, *45*, 18–42. https://www.researchgate.net/publication/271447352

McGrann, M. C., Tingley, M. W., Thorne, J. H., Elliott-Fisk, D. L., & McGrann, A. M. (2014). Heterogeneity in avian richness-environment relationships along the Pacific Crest Trail. *Avian Conservation and Ecology*, *9*(2). https://doi.org/10.5751/ACE-00695-090208

Merow, C., Smith, M. J., & Silander, J. A. (2013). A practical guide to MaxEnt for modeling species' distributions: What it does, and why inputs and settings matter. *Ecography*, *36*(10), 1058–1069. https://doi.org/10.1111/j.1600-0587.2013.07872.x

Mosher, B. A. (2011). *Avian Community Response to a Mountain Pine Beetle Epidemic*. Montana State University.

Muscarella, R., Galante, P. J., Soley-Guardia, M., Boria, R. A., Kass, J. M., Uriarte, M., & Anderson, R. P. (2014). ENMeval: An R package for conducting spatially independent evaluations and estimating optimal model complexity for Maxent ecological niche models . *Methods in Ecology and Evolution*, *5*(11), 1198–1205. https://doi.org/10.1111/2041-210x.12261

Odion, D. C., Moritz, M. A., Dellasala, D. A., Odion, D. C., Moritz, M. A., & Dellasala, D. A. (2010). Alternative community states maintained by fire in the Klamath Mountains, USA. *Journal of Ecology*, *98*(1), 96–105. https://doi.org/10.1111/j.

Ohmann, J. L., & Gregory, M. J. (2002). Predictive mapping of forest composition and structure with direct gradient analysis and nearest-neighbor imputation in coastal Oregon, U.S.A. *Canadian Journal of Forest Research*, *32*(4), 725–741. https://doi.org/10.1139/x02-011

Parker, T. J., Clancy, K. M., & Mathiasen, R. L. (2006). Interactions among fire, insects and pathogens in coniferous forests of the interior western United States and Canada. *Agricultural and Forest Entomology*, *8*(3), 167–189. https://doi.org/10.1111/j.1461-9563.2006.00305.x

Pase, C. P. (1982). Sierran Montane Conifer Forest. *Desert Plants*, *4*, 1–4.

Peterson, A. T., Martínez-Meyer, E., Araújo, M. B., Nakamura, M., Pearson, R. G., Anderson, R. P., & Soberón, J. (2011). *Ecological Niches and Geographic Distributions (MPB-49)*. Princeton University Press.

Phillips, L. F., & Schümm, S. A. (1987). Effect of regional

slope on drainage networks. *Geology, 15*(September), 813–816.

Plant, R. E. (2012). *Spatial data analysis in ecology and agriculture using R*. CRC Press LLC. https://doi.org/10.1201/b11769

Price, J. (2000). Modeling the potential impacts of climate change on the summer distributions of Massachusetts passerines. *Bird Observer, 28*(4), 224–230.

Pyle, P., & DeSante, D. F. (2003). Four-letter and six-letter alpha codes for birds recorded from the American Ornithologist's Union check-list area. *North American Bird Bander, 28*(2), 64–79.

Ralston, J., & Kirchman, J. J. (2013). Predicted range shifts in North American boreal forest birds and the effect of climate change on genetic diversity in blackpoll warblers (Setophaga striata). *Conservation Genetics, 14*(2), 543–555. https://doi.org/10.1007/s10592-012-0418-y

Ronen Kadmon, O. F. and A. D. (2004). Effect of Roadside Bias on the Accuracy of Predictive Maps Produced by Bioclimatic Models Author (s): Ronen Kadmon , Oren Farber and Avinoam Danin Published by : Wiley on behalf of the Ecological Society of America Stable URL : https://www.jstor.org/sta. *Ecological Applications, 14*(2), 401–413.

Russell, R. E., Saab, V. A., & Dudley, J. G. (2007). Habitat-Suitability Models for Cavity-Nesting Birds in a Postfire Landscape. *Journal of Wildlife Management, 71*(8), 2600–2611. https://doi.org/10.2193/2007-034

Saab, V. A., Latif, Q. S., Dresser, M. A., & Dudley, J. G. (2019). Woodpecker Nest Survival, Density, and a Pine Beetle Outbreak. *The Journal of Wildlife Management, 83*(6), 1387–1400. https://doi.org/10.1002/jwmg.21722

Schoenherr, A. A. (1992). *A natural history of California*. University of California Press.

Schrader-Patton, C., & Pywell, H. R. (2003). *Digital Aerial Sketchmapping*.

Seavy, N. E., Gardali, T., Golet, G. H., Jongsomjit, D., Kelsey, R., Matsumoto, S., Paine, S., & Stralberg, D. (2012). Integrating avian habitat distribution models into a conservation planning framework for the San Joaquin River, California, USA. *Natural Areas Journal, 32*(4), 420–426. https://doi.org/10.3375/043.032.0411

Service, U. F. (2010). *2010 Bark Beetle Damage*. https://www.fs.fed.us/foresthealth/applied-sciences/mapping-reporting/detection-surveys.shtml

Sousa, P. J. (1987). *Habitat Suitability Index Models: Hairy Woodpecker*. U.S. Department of the Interior, Fish and Wildlife Service.

St-Louis, V., Pidgeon, A. M., Kuemmerle, T., Sonnenschein, R., Radeloff, V. C., Clayton, M. K., Locke, B. A., Bash, D., & Hostert, P. (2014). Modelling avian biodiversity using raw, unclassified satellite imagery. *Philosophical Transactions of the Royal Society B: Biological Sciences, 369*(20130197), 1–10. https://doi.org/10.1098/rstb.2013.0197

Terziotti, S., & Archuleta, C.-A. M. (2020). Elevation-Derived Hydrography Acquisition Specifications. In *U.S. Geological Survey Techniques and Methods, book 11, chapter B11*. https://doi.org/ 10.3133/ tm11B11

Tingley, M. W., Monahan, W. B., Beissinger, S. R., & Moritz, C. (2009). Birds track their Grinnellian niche through a century of climate change. *Proceedings of the National Academy of Sciences, 106*(2), 1–7.

Tulloch, A. I. T., & Szabo, J. K. (2012). A behavioural ecology approach to understand volunteer surveying for citizen science datasets. *Emu, 112*(4), 313–325. https://doi.org/10.1071/MU12009

Turner, M. G., Arthaud, G. J., Engstrom, R. T., Hejl, S. J., Applications, S. E., Feb, N., Turner, M. G., Arthaud, G. J., & Engstrom, R. T. (2016). *Usefulness of Spatially Explicit Population Models in Land Management Liu , Susan Loeb and Kevin McKelvey Published by : Wiley Stable URL : http://www.jstor.org/stable/1942046 REFERENCES Linked references are available on JSTOR for this article : You may . 5*(1), 12–16.

U.S. Department Of Defense. (2007). *Global Positioning System Precise Posistioning Service Performance Standard* (Issue February).

U.S. Geological Survey. (2013). *USGS NED 1/3 arc-second 2013 1 x 1 degree ArcGrid*. USGS. http://nationalmap.gov/viewer.html

Weslien, J., & Schroeder, L. M. (1999). Population levels of bark beetles and associated insects in managed and unmanaged spruce stands. *Forest Ecology and Management, 115*(2–3), 267–275. https://doi.org/10.1016/S0378-1127(98)00405-8

Williams, P. L., & Koenig, W. D. (1980). Water Dependence of Birds in a Temperate Oak Woodland. *The Auk, 97*(2), 339–350. https://doi.org/10.1093/auk/97.2.339

Wilson, B. (2013). *Black-headed Grosbeak, Pheucticus melanocephalus*. Las Pilitas Nursery. https://www.laspilitas.com/California_birds/Cardinals/Black-Headed_Grosbeak/Black-headed_grosbeak.html#:~:text=The Black-headed Grosbeak is found throughout the western,our Black-headed grosbeak in the central plain states.

Zimmerman, J. L. (1997). Avian Community Responses to Fire, Grazing, and Drought in the Tallgrass Prairie. *Ecological Studies, 125*, 167–180.

Appendix

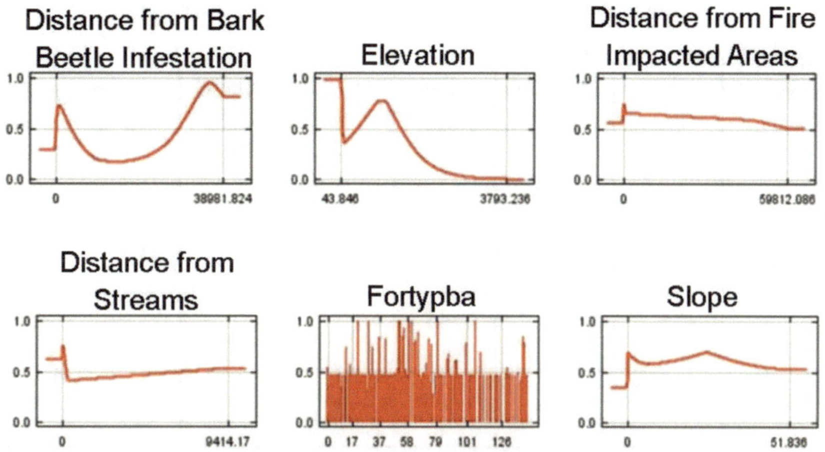

Appendix figure 1. BKHGRO response curves for the individual covariates.

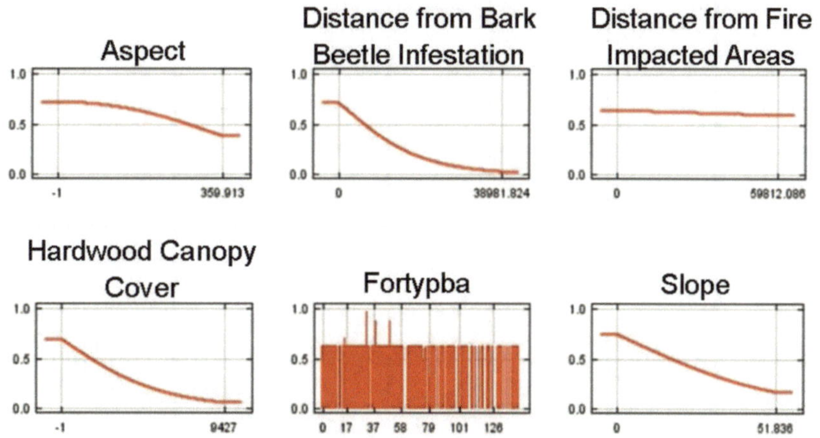

Appendix figure 2. HAIWOO response curves for the individual covariates.

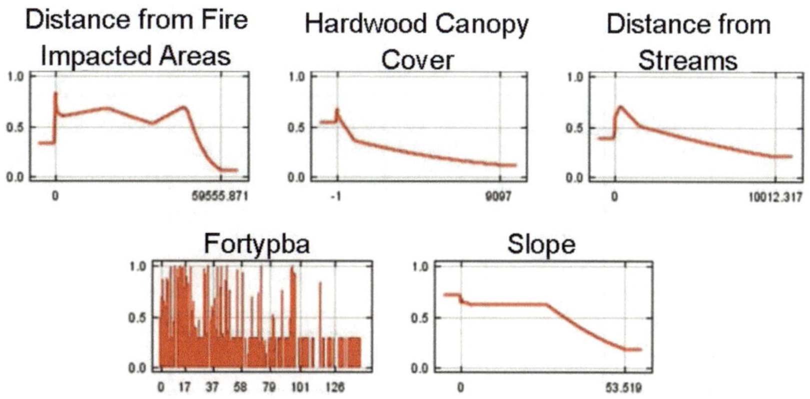

Appendix figure 3. YERWAR response curves for the individual covariates.

Table 4. Pearson correlation matrix produced in R (R code in the Appendix Section RCODE) of the correlation between environmental covariates. Covariates with a relationship above 0.7 were not used within the same model and prevent a potential reversal of the relationship within the model.

	Vegetation Class	Live Tree Density	Distance to Streams	Snag Density	Quadratic Mean Diameter of Dominant Tree Species	Hardwood Canopy Cover	Forest Type Based on the Basal Area of Dominant Tree Species
Vegetation Class	1.000	0.643	-0.315	0.157	0.638	-0.125	-0.223
Live Tree Density	0.643	1.000	-0.285	0.086	0.419	0.119	-0.194
Distance to Streams	-0.0315	-0.285	1.000	-0.152	-0.263	-0.100	0.122
Snag Density	0.157	0.086	-0.152	1.000	0.178	-0.122	-0.077
Quadratic Mean Diameter of Dominant Tree Species	0.638	0.419	-0.263	0.178	1.000	-0.162	-0.085
Hardwood Canopy Cover	-0.125	0.119	-0.100	-0.122	-0.162	1.000	0.390
Forest Type Based on the Basal Area of Fominant Tree Species	-0.223	-0.194	0.122	-0.077	-0.085	0.390	1.000
Distance to Fires	0.074	0.032	-0.248	-0.020	0.117	-0.031	0.015
Elevation	0.135	-0.017	-0.050	0.248	0.223	-0.653	-0.429
Conifer Canopy Cover	0.863	0.758	-0.354	0.150	0.624	-0.130	-0.236
Total Canopy Cover	0.763	0.780	-0.363	0.071	0.503	0.368	-0.041
Distance to Bark Beetle Infestations	-0.039	-0.049	0.078	-0.087	-0.051	-0.014	0.009
Aspect	-0.07	0.019	0.060	0.057	0.017	0.059	0.066
Slope	0.001	0.038	-0.354	0.143	0.096	0.262	0.170

Table 4 Continued

	Distnace to Fires	Elevation	Conifer Canopy Cover	Total Canopy Cover	Distance to Bark Beetle Infestations	Aspect	Slope
Vegetation Class	0.074	0.135	0.863	0.763	-0.039	-0.007	0.001
Live Tree Density	0.032	-0.017	0.758	0.780	-0.049	0.019	0.038
Distance to Streams	-0.248	-0.050	-0.354	-0.363	0.078	0.060	-0.354
Snag Density	-0.020	0.248	0.150	0.071	-0.087	0.057	0.143
Quadratic Mean Diameter of Dominant Tree Species	0.117	0.223	0.624	0.503	-0.051	0.017	0.096
Hardwood Canopy Cover	-0.031	-0.653	-0.130	0.368	-0.014	0.059	0.262
Forest Type Based on the Basal Area of Fominant Tree Species	0.015	-0.429	-0.236	-0.041	0.009	0.066	0.170
Distance to Fires	1.000	0.203	0.070	0.040	0.436	-0.044	0.328
Elevation	0.203	1.000	0.074	-0.220	0.071	-0.064	-0.046
Conifer Canopy Cover	0.070	0.074	1.000	0.858	-0.080	0.013	0.021
Total Canopy Cover	0.040	-0.220	0.858	1.000	-0.080	0.027	0.135
Distance to Bark Beetle Infestations	0.436	0.071	-0.080	-0.080	1.000	-0.065	0.040
Aspect	-0.044	-0.064	0.013	0.027	-0.065	1.000	0.039
Slope	0.328	-0.046	0.021	0.135	0.040	0.039	1.000

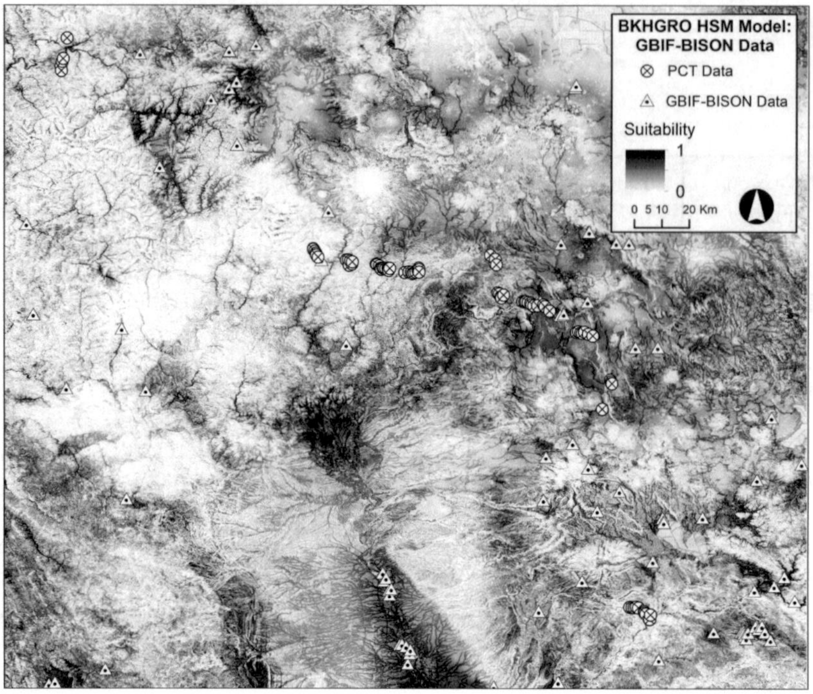

Appendix figure 4. BKHGRO habitat suitability model built from GBIF-BISON Data with regularization multiplier of 1.5.

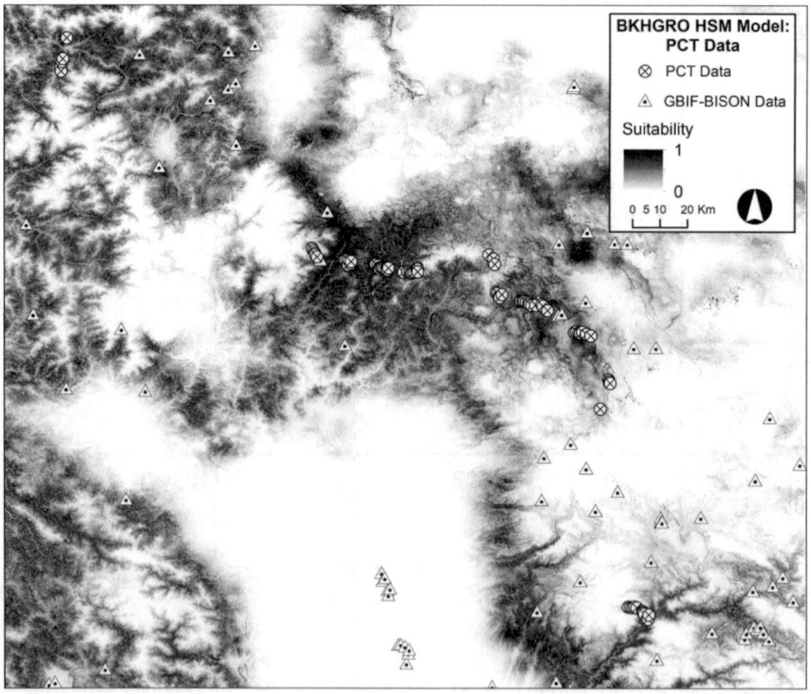

Appendix figure 5. BKHGRO habitat suitability model built from PCT Data with a regularization multiplier of 1.5.

Appendix figure 6. HAIWOO habitat suitability model built from the GBIF-BISON Dataset with a regularization multiplier of 5.5.

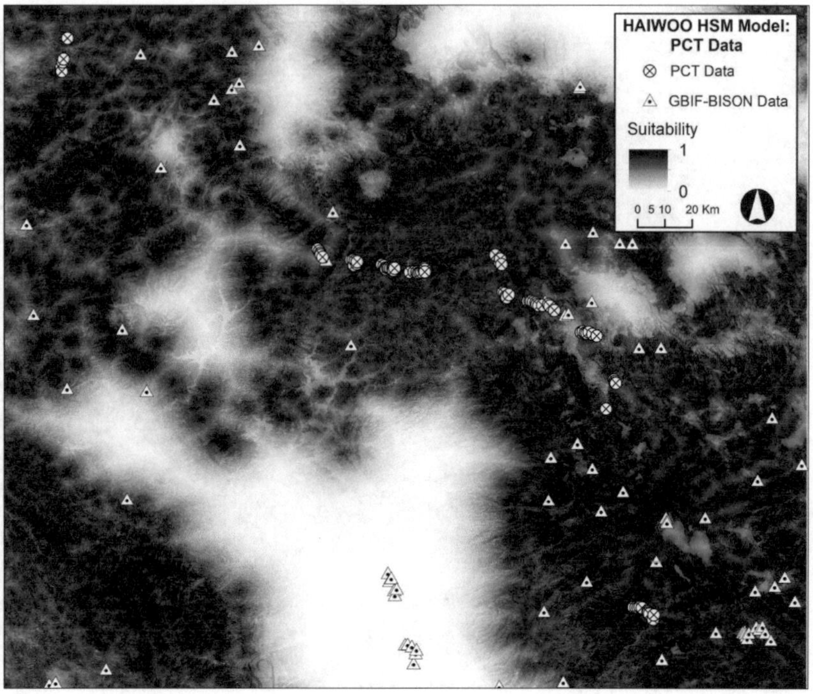

Appendix figure 7. HAIWOO habitat suitability model built from the PCT Data with a regularization multiplier of 5.5.

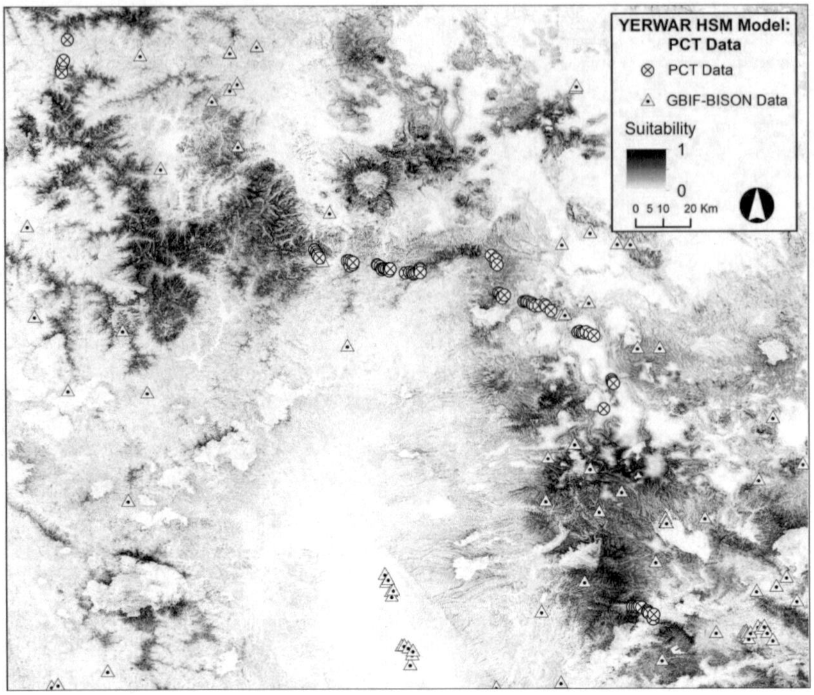

Appendix figure 8. YERWAR habitat suitability model built from the PCT Data with a regularization multiplier of 1.

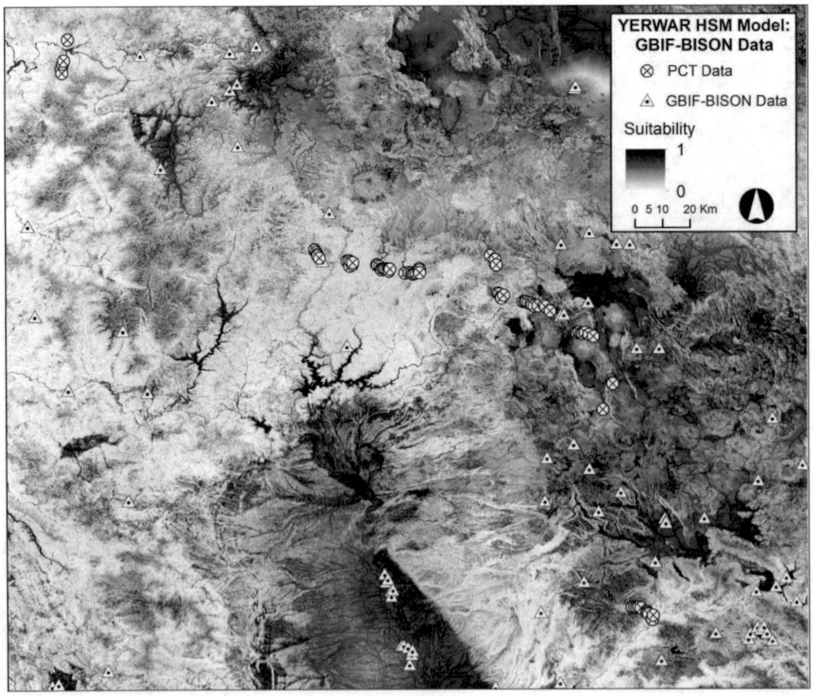

Appendix figure 9. YERWAR habitat suitability model built from the GBIF-BISON Data with a regularization multiplier of 1.

RSCRIPT

```
# Finding the correlation between environmental covariates
#adding data to R
dat<-(read.csv("", head=T))
attach(dat)
# Assessing correlations from the 12th column to the end
cor(dat[,12:ncol(dat)], use="pairwise.complete.obs",method="pearson")
#################################################################################
####
# R script to compare the suitability values for all the occurrence records for BKHGRO
# Import the suitability comparison csv file
BKHGRO=read.csv("")
plot(BKHGRO$Point, BKHGRO$maxent_R_4, pch=21, bg="black", xlab="Occurrence Record Number", ylab="Habitat Suitability")
# Finding the mean value of suitability
mean(BKHGRO$maxent_R_4)
#################################################################################
####
# R script to compare the suitability values for all the occurrence records for HAIWOO
# Import the suitability comparison csv file
HAIWOO=read.csv("")
plot(HAIWOO$Point, HAIWOO$MaxEntR3, pch=21, bg="black", xlab="Occurrence Record Number", ylab="Habitat Suitability")
# Finding the mean value of suitability
mean(HAIWOO$MaxEntR3)
#################################################################################
####
# R script to compare the suitability values for all the occurrence records for YERWAR
# Import the suitability comparison csv file
YERWAR=read.csv("")
plot(YERWAR$Point, YERWAR$MaxEntR3, pch=21, bg="black", xlab="Occurrence Record Number", ylab="Habitat Suitability")
# Finding the mean value of suitability
mean(YERWAR$MaxEntR3)
```

Python Script for calculating the standard error in the Aspect and Slope rasters

```
import os
import sys

# Open source spatial libraries
import shapely
import numpy
import gdal
import math
import random

# SpaPy libraries
from SpaPy import SpaBase
from SpaPy import SpaPlot
from SpaPy import SpaVectors
from SpaPy import SpaView
from SpaPy import SpaReferencing
from SpaPy import SpaDensify
```

```
from SpaPy import SpaView
from SpaPy import SpaRasters
from SpaPy import SpaTopo
from SpaPy import SpaRasterVectors
# set the input to the path where the original files are

#InputPath="C:\\Projects\\ProjectsPython\\HollsPaper\\Elevation.tif"
#InputPath="C:\\Projects\\ProjectsPython\\HollsPaper\\Sampled.tif"
InputPath1="C:\\Projects\\ProjectsPython\\HollsPaper\\ProjectedDEM_NoMask.tif"

OutputPath1="C:\\Projects\\ProjectsPython\\HollsPaper\\Slope1.tif"
OutputPath2="C:\\Projects\\ProjectsPython\\HollsPaper\\Slope2.tif"
OutputPath3="C:\\Projects\\ProjectsPython\\HollsPaper\\Temp.tif"

# Load the initial DEM
TheDEM=SpaRasters.SpaDatasetRaster()
TheDEM.Load(InputPath1)

WidthInPixels=TheDEM.GetWidthInPixels()
HeightInPixels=TheDEM.GetHeightInPixels()

# Create the base slope raster with no error
TheSlope=SpaTopo.Slope(TheDEM,OutputPath1)

TheSlope=SpaRasters.SpaDatasetRaster()
TheSlope.Load(OutputPath1)

# Setup the StdDev variables
SumOfSquares=0
N=0

# Loop over and over to improve the StdDev
Index=0
while (Index<1):

        # Load the DEM
        TheDEM3=SpaRasters.SpaDatasetRaster()
        TheDEM3.Load(InputPath1)
        TheBand=TheDEM3.GetBand(0)

        Row=0
        while (Row<HeightInPixels):
                Column=0
                while (Column<WidthInPixels):

                        Value=TheBand[Row][Column]

                        Random=numpy.random.normal(0,2.42)

                        Value+=Random
```

```
                TheBand[Row][Column]=Value

                Column+=1


            Row+=1

        TheDEM3.SetBands([TheBand])
        TheDEM3.Save("C:\\Projects\\ProjectsPython\\HollsPaper\\TempDEM.tif")

        SpaTopo.Slope("C:\\Projects\\ProjectsPython\\HollsPaper\\TempDEM.tif",OutputPath2)

        TheSlope2=SpaRasters.SpaDatasetRaster()
        TheSlope2.Load(OutputPath2)

        #
        TheBand1=TheSlope.GetBand(0)
        TheBand2=       TheSlope2.GetBand(0)

        Row=0
        while (Row<HeightInPixels):
            Column=0
            while (Column<WidthInPixels):

                Value1=TheBand1[Row][Column]
                Value2=TheBand2[Row][Column]

                if (Value1!=-9999) and (Value2!=-9999):
                    SumOfSquares+=(Value1-Value2)**2
                    N+=1

                Column+=1


            Row+=1


        Index+=1

StdDev=math.sqrt(SumOfSquares/N)
print(StdDev)
```

Eddylines: a metaphor for navigating Life's woes

Ann Nehaya Walsh

We sat on the fence and watched the blaze of wildfire, my seven siblings and I, ten yards away in the meadow we played in yards below our property.

The neighbor kids, the Halls, Cartwrights, and Wilsons, sat with us as firemen steadily tamed the fire to smoldering, blackened earth.

That was 1972 in Carmel Valley, California, where wildfire was an occasional occurrence from late August into October during my first fifteen years. Save for the meadow being ablaze, most fires were distant; plumes of grey smoke rising as a huge billowy column further down the valley, or in the coastal range across the valley from our rural neighborhood.

We left Carmel Valley when I was fifteen for another culture, geography, and climate.

Eventually, after living in other far-off ports, I returned to live in California more than forty years later.

That was autumn 2018, when wildfires in the US western states and Canada's British Columbia were the worst anyone had known.

Autumn 2019 was worse than before. This horror was compounded with Australia's dry, red-soiled Land, ravaged by fire that decimated some of the rarest creatures on Earth.

I'd lived in Australia when I was 25. The experiences of that odd land had touched me deeply. It wasn't until Australia went up in smoke that I began to feel and grieve what had and was currently happening on my native soil.

Now, autumn 2020, the wildfire season has been horrific in scope and impossible to grapple with.

I've learned wildfire makes its own weather patterns; lightning, cyclones of fire, and fireballs spitting as far as eight miles out from the center, igniting more fire amongst dry grasses, trees, and soil. Wildfire can also make thunderstorms and rain.

Even though I've been deeply blessed to have, so far, lost nothing to fire, I've listened to people's stories of fighting fire, of homes and towns being destroyed, and the absolute devastation to people's hearts and souls.

I have endured long days and weeks of air so choked with smoke that even avid walkers, such as I, wear masks for brief, small walks throughout the day.

Throughout these experiences, moist memories rose of kayaking British Columbia's gulf islands and Washington State's San Juan islands with my then husband Keith Fredrikson.

Keith and I kayaked often, and I learned to traverse eddylines in a hardy, sea-faring kayak.

Eddylines are two currents that abut each other, causing a harrowing current that with patience and skill can be crossed.

*

The air was chilly and damp. Low, gray fog lifted slowly from a small headland as we kayaked past.

Cold, gray-blue water swirled into stronger currents beneath my paddle's blades.

"Watch the current," Keith called from my right.

Nodding, I adjusted my grip on the paddle's rod and continued paddling rhythmically.

Blade in, slush, pull back and out. Seawater dripping from the blade sprayed my face. Other blade in; slush, pull back… suddenly the blade was torn askew underwater. Out came the blade. Other blade in. No beat missed. A jagged entry. Pulled back the paddle. The blade skirtled sharply left. Lost momentum. I pulled back again, out, and looked ahead.

The current jutted up, forming a small, white line of water streaming away from the island's head.

I wasn't fooled by the smallness of the white water. "Eddyline!" Keith called. "Keep a blade in the water." Glancing at him, I nodded briefly, glad for his guidance.

Sinking my hips deeper into my kayak's plastic seat, I kept the bow directed towards the eddyline and was funneled forward.

*

Now, the 2020 wildfire season has been compounded by Covid; which has spread like waves of fire throughout our species.

In light of this, I must ask myself, again, how have I participated so that this has come about? For, I feel and see a direct correlation between our ways of living and using Earth these past decades, century, and longer that has made the firestorm of Covid-19.

As for my participation…

True; I've been purposely without a car for two+ years; I have composted and recycled for 30+ years; eaten and grown organic foods for donkey years; and purchase foods from Humboldt County and Northern California shoreline…

It's also true that I happily buy foods grown or fished thousands of miles from my neighborhood; would book a ticket and hop on a plane taking me to my parents in San Diego, Ireland, and Maui

in a minute; I drove a vehicle copiously for decades and have, at times, felt socially pressured to seriously consider buying a car so I've more options for employment; and I may not remain in North America for permanent settlement.

Many of us have paths in Life that are enthralling and fulfilling, which have us traveling, teaching, researching, and forming fresh protocols for professional conduct in ports far from our own neighborhoods.

It's kindly true, also, to include the visions of explorations we wish for those we love who have held back on going to the Trinidads, Istanbuls, and Tanzanias of this beautiful, bountiful World… because… they can.

However, this belief of travelling hither and thither in our counties, States, and further forms an eddyline with our awareness that we must deeply change our daily ways of living because our treatment of Earth is bringing Humans and Earth to our knees.

*

Earth has been an extraordinarily gracious Hostess to us.

Many of my colleagues and I know/feel Earth's intelligence and believe She resonates at a frequency higher than Humans.

Earth can swallow us whole in a matter of breaths. She could regurgitate whatever She wants and, with time, bring Life to Herself anew.

I sense Earth is well aware of the taxation being done to Her, and is willing to wait things out till we bring ourselves to our own knees, and our Minds humble into our Hearts.

For all my experiences, Divine communiques and guidances, I do believe a stout few can shift Consciousness for many.

How can this be done? And how can this come about?

Love—truly, tenderly, without bounds of judgement, shame, nor restriction—One's Own Dear Self.

Love—especially in the Dark Hour—however this hour shows itself; with difficult aspects of yourself; with a natural catastrophe decimating your home, land, trees, and animals; with a troubling colleague; a disturbing neighbor; the too busy friend or spouse; the loss of something or one Dear; with governmental Leaders; with Covid-19.

Feel the anger. Feel the grief and pain. Feel compassion for yourself; compassion for that, or who, you hate.

Love these things, too. And be thankful for what is good and steady.

For all your ills will, too, pass.

*

The current roughened.

Wind pushed deep against my stern.

Glad I was, for a heavily packed bow and stern with a fortnight of food and water.Keep a blade in, I chanted, terrified of capsizing.

Keith had practiced "Eskimo rolls" in a pool, then later Lake Padden as I watched how he did it.

I hadn't the guts to purposely capsize; sprayskirt swaddled and latched to my cockpit as my body and kayak met water, with pole in hand rudder some manuvers underwater and, Voila, Pop!

Up to surface, water flooding off everything.

No thank you.

Instead I practiced ripping my sprayskirt free of its latched sealing. If I capsized I trusted my slithering and swimming finesse more than purposely staying captive till, maybe, I surfaced.

Grey-blue water sucked and pushed at my blades in jagged swirls.

The water was cold and deep. Chilly wind sunk beneath my raincoat.

Don't think. Blade in. Pull back. Out. Exhale. Blade in.

I churned through the wide eddyline. Controlled Fire Mind.

Ache built in my shoulders. My knees pressed the hunnells deep. They would be bruised later.

I shortened the reaches of my paddle. All that existed was the fierce, contorted current. Blade plunged hard and deep, slush. Back. Up. Seaspray drenched me. Blade plunged.

Sharp wind cut against my neck. Heat fumed up beneath my coat.

Keith was lengths ahead. I smiled, wryly. It was how we walked, lived. I slowed. He quickened.

My shoulders burned from battling the current. I persevered. Blade in, pulled deep, then.... Nothing. No resistance. The current carried my kayak as I thrust a blade in, not trusting to break the quick, stout rhythm.

Whoosh. I was carried swiftly away from the eddyline, caught Keith's eye and shook my head as if to say, "Damn. Whew!"

He smiled, noddingly. We kayaked abreast, leisurely and were carried swiftly around a small headland and began to seek camp for the night.

*

I believe as each of us continues to do our part as we can, we will come through this stalwart eddyline that we have been in for some time and have more to traverse until we come to good, safe shore.

Surely, the Love I feel for myself, Earth, and Humanity is my prayer and pervades my entire presence. This prayer—this "kayak and paddle"—I offer in every situation.

Will the great trees, animals, and birds of western North America and Australia replenish themselves...? I pray enough will. This trust entwines with loving, yes, loving these recent horrors and rips to our Psyches... because I can.

We can.

Stepping from Behind the Redwood Curtain: Using a Cultural Wealth Approach to Support Study Abroad at Humboldt State University

Alison R. Holmes (Humboldt State University), Sam Lipiec (Humboldt State University), Ileanna Spoelstra (Humboldt State University)

Abstract

Studying Abroad is considered a "high impact practice" because it is one of the most transformational experiences offered in a college setting. Humboldt State University (HSU) has an excellent record of sending a diverse range of young people abroad (including first generation, low income, and minoritized or under-represented populations often known as URMs) when compared to other universities. However, as our demographics change and campus staffing and unit responsibilities shift, we need to constantly review our processes to ensure best practice in support of student success. This project set out to assess HSU's current systems and the connection between curricular learning outcomes and staff advising and support for study abroad. Our conclusions point to a number of observations and practical recommendations broadly guided by Tara Yosso's six-part Cultural Wealth Model. Specifically, we have found that Yosso's approach—which she only applies to students of color—should be applied to all students as it would benefit anyone seeking this experience. Our findings, using interviews with relevant faculty and staff as well as study abroad data collected at HSU, suggest a need to better connect academic and student support and advising services and the creation of a more collaborative, strength-based (i.e. Yosso inspired) support network across campus, as we promote study abroad opportunities and help students prepare for, and return from, their experience overseas. The goal should be the addition of a global dimension to the "cultural wealth" of all our students.

Keywords: Study abroad, cultural wealth, student affairs/curricular integration

Humboldt State University (HSU) is a small, remote, rural campus located on the far northern coast of California. Historically, HSU started as a normal college to train teachers and nurses, but eventually joined the California State University system. Today HSU is a liberal arts university with just under 7,000 students, 48 majors and 61 minors divided among three colleges: Arts Humanities and Social Sciences (CAHSS), Natural Resources and Sciences (CNRS) and the Professional Studies (CPS).

In an effort to assess the processes and potential roadblocks that students seeking to go abroad may experience, this project examined Humboldt State University's (HSU) study abroad student data and interviewed six staff members directly relevant to the study abroad process. This included staff from the areas of academic affairs, the cultural centers, program advising and financial aid. Through this project, it became clear that more could be done to better understand the needs of our student populations and to connect the existing advising efforts on campus. A cultural wealth model argues that higher education should do more to bring the diverse perspectives, "knowledge, skills, abilities, and contacts possessed and utilized" by different cultural or identity groups into academic spaces (Yosso, 2005). However, given that study abroad is not owned by any specific academic area but brings a more global perspective to the entire campus, there is work to be done to bring together the curricular and co-curricular support for this "high impact practice," delineated by scholars such as George Kuh, as an experience that makes a significant difference to overall student success (Kuh, 2008). Institutions of higher education have traditionally focused on the academic or curricular aspects of campus life often to the exclusion of other aspects of student well-being. However, as student demographics change, the mission of equity and accessibility must increasingly influence our efforts to improve retention and graduation. Our goal is to show that a holistic approach to the relationship between academic goals, student support and identity, through extra- or non-curricular activities, is a vital factor in student success and fundamental to any advising support infrastructure. This paper argues that HSU's evolved career curriculum model offers a template for study abroad by undertaking more proactive advising to engage students.

In some respects, HSU was ahead of the curve from the founding of what today is called the Native American Center and the creation of a student-run Multicultural Center (MCC) twenty-five years ago. Less than ten years ago the university established the Centers for Academic Excellence (CAEs) (e.g. the African American Center for Academic Excellence and El Centro Académico Cultural) as a further effort to address the gap identified between academic and co-curricular support. However, relatively little was done at that time to evolve what could be seen as more traditional campus units such as the Career Center or, more relevant in this case, the study abroad office, now called the Center for International Programs. Both of these units maintained the more traditional profile of waiting for students to take up their services rather than reaching out directly to the increasingly diverse student population. That said, eight years ago, the College of Arts, Humanities and Social Sciences created a committee specifically charged with the task of reimagining, scaffolding and integrating the career curriculum into both the academic programs and co-curricular areas. Since then, this innovative approach has spread across the university.

In Kuh's terms, study abroad is recognized as one of several high impact practices, but significant racial and ethnic disparities remain. According to the National Association of International Educators (NAFSA) nearly 71% of students who study abroad identify themselves as white. While this is often assumed to be true across the country, this national data stands in contrast to HSU where a relatively small number of students go abroad, but they are from a wide range of backgrounds. Specific data pulled from HSU's Office of the Registrar indicates that over the period covering Fall 2014 through Summer 2018, 579 HSU students studied abroad. This breaks down by the university's three colleges as follows: 330 from Colleges of Arts, Humanities and Social Sciences (CAHSS); 131 from Colleges of Natural Resources and Sciences (CNRS); 96 from College of Professional Studies (CPS); and 22 undeclared. The largest major sending students abroad by a significant margin is International Studies (INTL), a small interdisciplinary program that sent 86 students abroad in this period (representing 15% of all HSU students who studied abroad and over 25% of the students from CAHSS). More impressive is the fact that over 50% of all students going abroad from HSU were first generation and 43% identified as under-represented minorities.

Tasha Willis, among various scholars, has offered some recommendations to help students of color make the most of their study abroad experience by taking advantage of (1) travel peers, (2) campus climate, and (3) critical reflection (Willis, 2015; Buffie, 2019). However, HSU's experience with similar students suggests that intensive advising and support before and after a study abroad experience, including coursework that focuses on reflection, may also help address the needs of a broader range of students. Similarly, it suggests that the cultural centers could potentially have more of a role both in offering study abroad programming as well as support both pre and post a student's foreign experience. Such conclusions are supported by the approach taken by the International Studies program, which requires a study abroad as part of the major. This ensures intensive advising and support for all students as well as a presumption of success that could be broadly identified as following the cultural wealth model. The fact that this program's demographics reflect those of the university as a whole (i.e. not skewed to white or wealthy students, which one might expect based on the national statistics) and all students in the program successfully complete a study abroad.

The International Studies program's intentional support network could also help to explain student choices in terms of the programs they choose to use for study abroad. For example, again using the Registrar's data from 2014-18, HSU's faculty-led programs are consistently a first choice for students and assures them they will not only have travel peers, but faculty who are known to them (249 students). Similarly, HSU students also regularly choose California State International Programs (CSUIP) where they will at least be familiar with the system and faculty style (110 students). By extension, it seems logical that other programs have less take-up by HSU students but remain fairly evenly balanced and include bilateral programs (110 students) and various consortia of universities (100 students). This leaves the smallest group as those choosing organizations that students apply to directly (27 students) with some anecdotal evidence that students seek these out for a specific location or type of programming.

Higher education has long operated on models of learning such as Abraham Maslow's hierarchy of needs or, more recently, Benjamin Bloom's taxonomy, as attempts to apply theories of learning to educational pedagogies. Traditionally, most academics have focused almost exclusively on Bloom's Cognitive Domain that covers the area of mental skills and knowledge acquisition (Bloom et al, 1956). While many have attempted to focus on the higher levels of this domain (e.g. evaluating and creating vs rote facts and remembering) there has been relatively little effort to include Bloom's other two domains: Psychomotor (mainly physical and deemed more appropriate to manual/skills learning than higher education); and, more relevant here, the Affective Domain (which includes emotional learning, attitudes and awareness of self) (Anderson et al, 2001). Though updated in the 1990s, Bloom et al have been critiqued by scholars, such as Tara Yosso, who identified gaps in this cognitive-centric view - particularly for students of color. Yosso therefore developed a Cultural Wealth model focusing almost exclusively on students who need a more holistic approach and the identification of what is termed the six forms of "cultural capital." Yosso's six forms of cultural capital are aspirational, linguistic, familial, social, navigational, and resistance. Each form represents the unique qualities students of color embody and develop while attending college. Yosso argues that cultural capital helps enrich a student's learning experience (Yosso, 2005). The result of this approach and similar critiques has been to help campuses reflect on the connection between student identity and what they effectively bring with them to campus and how that can be better recognized and effectively honored. Furthermore, such approaches focus on ways in which academic and support units—the curricular and the co-curricular—could better meet each student "where they are."

In that spirit, HSU's career center began to work directly with faculty on ways to extend its offering through major programs, as well as more direct services like the CAEs and other student programming spaces such as library *Skillshops*. The slow, but organic, evolution of this holistic approach, now reaches deep into a huge range of academic programs and helps students connect their real-world career interests to their scholarly pursuits and to better articulate and translate the skills they have gained to their plans after college. The career center staff have found that this approach provided a huge step forward in terms of reaching more of the diverse student body and a more coherent and collaborative effort between staff and faculty. However, to date, this integrative approach has not been extended to other areas—including study abroad. When interviewed, staff in all the CAEs regularly expressed a need for promotional and informational materials that are "welcoming and representative" of different groups as suggested by scholar Whitney Schulze.

Furthermore, and again in keeping with Schulze, the CAEs felt the need for any such materials to be designed *in conjunction* with other units on campus (Schulze, 2016). HSU staff working in these co-curricular spaces regularly identified what they perceived as their own lack of knowledge to be a potential issue and voiced an interest in having more information and training, or, at the very least, some kind of fact sheet they could offer students. They also expressed a desire for a more concrete and practical connection with the study abroad office, in much the same way other services or units on campus, not only careers but the advising center and the library, have developed plans to include the CAEs in their outreach. Interestingly, in the interviews conducted for this project, faculty in the colleges and majors that do not traditionally send many students abroad echoed these same sentiments. This makes the contrast all the more distinct from the sentiment of the study abroad office, who felt they were offering information and materials that all students were welcome to access—if they came to them—and were often disappointed that more students did not take advantage of the resources they felt to be freely available.

Data gathered by the faculty of the International Studies program as part of their leadership of HSU's International Education Week clearly suggests that more students are interested in studying abroad than reflected in the numbers who end up going (Holmes, 2014-2020). Since 2016, 100% of students who offered feedback at this week-long campus event have indicated that they would like to study abroad. However, when asked why they might not, their concerns were twofold. The issue most commonly listed by far was financial (or lack of knowledge on how to plan/using financial aid). Secondly, they were concerned about time to graduation (and/or what they felt to be the complexity of the process/lack of advisor support). This information all seems to indicate a significant disconnect between what *students* believe to be the obstacles and the perspectives of those advising students at key points in their college career.

A number of practical steps suggest themselves. For example, a relatively easy first step would be a set of general information and inclusive fact sheets could be designed for CAE staff as well as faculty across all colleges. These could be very general for staff in different offices around campus, but they could also be tailored to the university's three colleges or even to a specific major. This would not only help faculty and staff feel more confident they are giving correct information, but they would also begin to counter the disturbingly common problem of students believing it is "too expensive" or will delay their time to graduation. While there is generally a need for more financial awareness among many groups on campus—and particularly among students of color—all such information should be sure to include early and specific information on how study abroad can be organized and planned so that it need not be the burden that many students fear. Similarly, materials and events that showcase the fact that HSU, contrary to common mythology, already sends a range of students abroad, could include testimonials, panel discussions, and student ambassadors to conduct outreach to specific CAEs. Following the career curriculum model, faculty champions in each college could be created to ensure a stronger connection between the staff and faculty and begin to integrate this high impact practice into more majors and programs. None of these initiatives are difficult, but such relatively simple steps would go a long way to creating the necessary campus climate and advising support for all students, and particularly students of color.

Yosso focused on students of color, but this research suggests there is much to be gained by taking what could be called a broad cultural wealth approach to meet all students where they are in the area of study abroad at HSU. A more coordinated and collaborative approach between staff and faculty would not only offer more students the opportunity of a lifetime and the ability to develop their global cultural wealth, but such an approach would also bring a much-needed global awareness and wealth back to the campus as a whole.

References

Anderson, L.W., Krathwohl, D.R., Airasian, P.W., Cruikshank, K.A., Mayer, R.E., Pintrich, P.R., Raths, J., Wittrock, M.C. (2001). *A Taxonomy for Learning, Teaching and Assessing: A revision of Bloom's Taxonomy of Educational Objectives.* New York: Pearson: Allyn & Bacon.

Bloom, B.S. (Ed.). Engelhart, M.D., Furst, E.J., Hill, W.H., Krathwohl, D.R. (1956). *Taxonomy of Educational Objectives, Handbook I:The Cognitive Domain.* New York: David McKay Co Inc.

Holmes, Alison (2020, 2018, 2017, 2016, 2015, 2014). Unpublished feedback/survey data from Humboldt State University's International Education Week.

Holmes, Alison and Collins, Loren. (2017). "Career Planning and Curriculum Integration: millennials on the 'lost' coast" Career Curriculum Working Paper Series, HSU Press. *Online.*

Kuh, George W. 2008. *High-impact educational practices: What they are, who has access to them, and why they matter.* American Association of Colleges and Universities.

Longmire-Avital, Buffie. (2019) "Establishing Equity in Study Abroad Experiences: Three Recommendations". (December 17) https://www.centerforengagedlearning.org/establishing-equity-in-study-abroad-experiences.

Maslow, A. H. (1943). "A Theory of Human Motivation". *Psychological Review, 50(4)*, 370-96.

NAFSA. n.d. "Trends in U.S. Study Abroad". Accessed December 11, 2019, https://www.nafsa.org/policy-and-advocacy/policy-resources/trends-us-study-abroad.

Schulze, Whitney L. (2016) "Best Practices for Increasing

Diversity in Study Abroad: A Manual for Small Private Co-Ed Universities in the United States," *Projects and Capstones*. https://repository.usfca.edu/capstone/347/

Willis, Tasha Y. 2015. "And Still We Rise…": Microaggressions and Intersectionality in the Study Abroad Experiences of Black Women. *Frontiers: The Interdisciplinary Journal of Study Abroad* 26, no. XXVI (Fall): 209-230.

Yosso, T.J. (2005). "Whose culture has capital?". *Race, Ethnicity and Education*, 8(1), pp. 69–91.

Cannabis Tourism and the Community: Resident Attitudes in Humboldt County, California

Ara Pachmayer (Humboldt State University), Reed Switzer (Humboldt State University), Shawn Reilly (Humboldt State University)

Abstract

Resident attitudes towards cannabis tourism in Humboldt County, California were investigated in this exploratory study. The primary purpose was to understand if the variables used to explain resident attitudes towards tourism would show similar patterns when applied to cannabis tourism. The study included a questionnaire distributed to adult residents of Humboldt County and was conducted between March and August 2017, just prior to recreational cannabis sales beginning in California. Correlations and crosstabs were performed on the data. While personal benefit was found to have an impact on resident attitudes towards cannabis tourism, other variables indicate mixed support which suggests that some variables may behave differently when considering cannabis tourism. Managerial implications are also considered.

Keywords: resident attitudes, community, tourism, cannabis tourism

Introduction

Cannabis tourism has been defined as "purchasing with the intent to consume marijuana products while temporarily traveling away from one's normal place of work or residence" (Taylor, 2019, p. 6). In contrast, others have determined that the tourist does not need to have the intent to consume cannabis to be participating in cannabis tourism but can simply be interested in learning more about cannabis (Giraudo, 2019). Cannabis tourism might also include visits to dispensaries or facilities involved in cannabis processing, attending cannabis friendly retreats, and educational courses such as culinary cannabis classes.

In 2016, the passing of Proposition 64 allowed for the sale and taxation of recreational cannabis to begin in California on January 1, 2018. Proposition 64 opened the door for legal cannabis tourism in California and Humboldt County (Houston, 2016). Due to the international reputation of cannabis production in Humboldt County, many in the Humboldt County tourism industry saw this as an opportunity for Humboldt County to become the "Napa Valley of cannabis tourism" ("Get ready for marijuana legalization, or get ready to lose," 2014). Cannabis tourism is a potential complement to Humboldt County's existing tourism resources, which include scenic coastlines, rural beaches, redwood forests, and a variety of local businesses providing unique products and experiences.

Looking to the experiences of other US states, according to the Colorado Tourism Office, the number of out of state visitors to Colorado who reported that they were more likely to visit the state due to cannabis has increased by 10% since the legalization of cannabis ("Economic effects," 2016). Another study found that after legalization in 2014, tourists made up 44% of recreational sales in Denver and 90% of recreational sales in mountain tourist areas (Light et al., 2015). The potential tax revenue for local communities is tremendous.

Claesgen and Kraft (2018) indicate tourism as a key component related to the cannabis industry. Beyond being the home of the redwoods, Humboldt County often promotes the wealth of locally grown and/or produced products as a selling point for visitors to the area ("Travel info for the Redwoods," n.d.). According to Claesgen and Kraft, "it is believed cannabis tourism could help increase the demand for . . . Humboldt-branded products" (p. 9). Many communities throughout the county and other legal states are actively pursuing cannabis tourism. Despite this being seen as a good opportunity for Humboldt County and other communities in legal states, few studies have focused on how residents feel about cannabis tourism (Kang et al., 2016b).

Understanding resident attitudes towards tourism is a critical piece of tourism development and growth (Sharpley, 2014). The current study attempted to capture the attitudes of the general public across Humboldt County rather than focusing on a specific population. Another similar study is not known to the authors at this time.

The purpose of the study was to explore if the variables used to explain resident attitudes towards tourism would show similar patterns when applied to cannabis tourism in Humboldt County, California. Given that legal cannabis tourism is a relatively new research topic, questions to guide the study were developed utilizing past research in the area of resident attitudes towards tourism (Andereck et al., 2005; Gursoy & Rutherford, 2004; Látková & Vogt, 2012; McGehee & Andereck, 2004; Perdue et al., 1990). The study focused on two research questions including:

1. Is there a relationship between age, length of residency, and personal benefit from, and attitudes towards, cannabis tourism?
2. Do attitudes towards cannabis tourism differ in terms of the level of knowledge about the tourism industry and personal benefit from tourism?

Methods

In advance of research taking place, the protocol for data collection was approved by Humboldt State University's Institutional Review Board. Data collection involved a paper-based and online questionnaire. The questionnaire was developed to assess resident attitudes towards tourism in general with several items aimed more specifically towards cannabis tourism. Respondents were asked a series of Likert scale items regarding attitudes towards cannabis tourism. Multiple items to assess attitudes toward cannabis tourism were adapted from Kang, Miller, and O'Leary (2016a).

In-person surveying of adult residents occurred at popular spots throughout the county in March and April 2017. Diverse sites were selected to capture residents including shopping centers, downtown districts, parks, restaurants, and cafes (Table 1). Permission to survey was sought from the appropriate people in advance of surveying. For onsite surveying, students from the Spring 2017 REC365 Travel Industry Management class at Humboldt State University were trained in random sampling and the survey distribution process and completed all in-person surveying.

An online questionnaire was made available for residents to complete between May and August 2017 to capture residents living in remote areas. Distributed through convenience and snowball sampling, a link to the online questionnaire was emailed to roughly 50 people who were asked to further distribute the link. The link to the online questionnaire was also distributed through local media outlets and press releases. In total, between onsite and online surveying, 806 individuals responded to the cannabis tourism attitude items on the survey. A majority of participants responded to the online survey which was a convenience and not a random sample. Therefore, the results apply only to people who responded to the questionnaire.

Literature

While tourism may provide an economic advantage for communities, it may also lead to negative impacts on the community and its residents. Without resident support, tourism may not be accepted in a community, and the tour-

Table 1. Survey sites and months for onsite surveying.

Survey Sites	**Months**
Mad River County Park, Arcata	March 2017
Rohnerville Park, Fortuna	March 2017
North Coast Co-op, Eureka	March 2017
Eureka Natural Foods, Eureka	March 201
Bayshore Mall, Eureka	March 2017 – April 2017
Union Town Shopping Center, Arcata	March 2017 – April 2017
Old Town, Eureka	March 2017 – April 2017
Arcata Plaza, Arcata	March 2017 – April 2017
College Cove, Trinidad	April 2017
Samoa Dunes Recreation Area, Samoa	April 2017
Woodley Island, Eureka	April 2017
Rays Grocery Store, Fortuna	April 2017
Redwood Curtain Brewery, Arcata	April 2017
Mad River Brewery, Blue Lake	April 2017
Starbucks, McKinleyville	April 2017
Cher-Ae Heights Casino, Trinidad	April 2017

ism resources that originally brought visitors to a destination may be changed beyond recognition (Martin, 1995). Many studies have sought to better understand resident perceptions of the impacts of tourism (Andereck et al., 2005; Haley et al., 2005; Williams & Lawson, 2001). Beyond understanding perceived impacts, stakeholders should also know the factors that influence resident attitudes towards tourism. Gursoy, Jurowski and Uysal (2002) state that:

> "while success in the industry depends upon attractions and services, it requires the hospitality of local residents . . . Understanding local reaction and the factors that influence these attitudes is essential in achieving the goal of favorable support for tourism development." (p. 80).

There are a host of variables that have been studied to understand their influence on resident attitude towards tourism including the level of tourism development in a community, economic dependency on tourism, community attachment, how much contact an individual has with tourists, knowledge of the industry, and demographic characteristics (Gursoy & Rutherford, 2004; Hao et al., 2011; Jurowski & Gursoy, 2004; Sinclair-Maragh, 2017; Williams & Lawson, 2001). As noted above, the current study focused on resident attitudes towards cannabis tourism based on age, length of residency, knowledge of the tourism industry, and personal benefit from tourism and cannabis tourism.

In general, when considering resident attitudes and demographic characteristics such as age, length of residency, gender and level of education or income, no relationships have been found and often results have been conflicting (Cui & Ryan, 2011; Hao et al., 2011; Madrigal, 1993; McGehee & Andereck, 2004; Perdue et al., 1990; Sinclair-Maragh, 2017). For the current study, demographic statistics were collected to develop a profile of the community under study but also to understand any influence these variables had on resident attitudes towards cannabis tourism.

In past research, age has not been shown to have an impact on resident attitudes towards tourism with few exceptions (McGehee & Andereck, 2004; Sinclair-Maragh, 2017; Weaver & Lawton, 2001). Sinclair-Maragh investigated the demographic profile of residents and support for tourism in Jamaica. They found that younger residents had more support towards tourism. McGehee and Andereck explored a variety of factors that predict attitudes towards tourism. In terms of age, when the authors controlled for personal benefit they found that older residents were more supportive of tourism concluding that age did predict attitudes towards tourism. Age was also found to be a factor in determining attitudes towards tourism in Weaver and Lawton with older residents having more negative attitudes towards tourism. However, the authors found that older residents also had longer term residency in the community which may have been a factor in their negative attitudes (Weaver & Lawton). In light of the mixed results from past research, age was considered in the current study given the changing stigma associated with cannabis.

Length of residency is another example of a demographic characteristic that has been included in several studies considering resident attitudes towards tourism (Andereck et al., 2005; Hao et al., 2010; McCool & Martin, 1994; Sinclair-Maragh, 2017). Similar to age, in past studies conflicting relationships or no relationships have been found when considering the impact of length of residency on resident attitudes. Andereck et al (2005) conducted a study in Arizona to examine resident perceptions of tourism impacts. They tested the relationship between resident attitudes toward tourism and community attachment determined by length of residency along with several other demographic variables. The authors posited that those who had grown up in the community, or those with longer residence in the community, would have greater attachment to the community and as a result would perceive fewer positive impacts from tourism. In contrast to other studies, the results showed a weak and insignificant relationship (Andereck et al., 2005). Sinclair-Maragh (2017) hypothesized that length of residency would not impact support for tourism development. While it was found that residents with more than 20 years tenure in their community supported tourism development, the results were not statistically significant and the author concluded that length of residency did not impact support. Additional research found conflicting results when exploring length of residency and attitudes toward tourism. Hao et al. (2010) explored variables that influenced resident attitudes toward tourism in a coastal resort community in North Carolina. In the community under study, Hao et al. found that people who had longer lengths of residency also had more positive attitudes toward tourism.

Like age and length of residency, no clear relationship between the level of knowledge of the tourism industry and resident attitudes can be found across several studies (Andereck et al., 2005; Látková & Vogt, 2012). Látková and Vogt considered the influence of subjective knowledge on residents' attitudes towards existing and future tourism development. When controlling for personal benefit from tourism, the authors found subjective knowledge about tourism did not predict residents' perceptions of positive or negative impacts of tourism. Andereck et al. found contrasting results in their study on residents' perceptions of the impacts of tourism. The study found that residents who indicated greater knowledge about tourism were more positive towards tourism. Due to the newness of legal cannabis tourism in the state, it was determined to ask participants to describe their level of knowledge of the tourism industry.

In contrast to the other variables examined, when considering personal benefit from tourism and resident attitudes, a clear pattern can be found. Previous studies considering tourism, not cannabis tourism, have shown that individuals who indicate they benefit from tourism generally have more positive attitudes towards tourism (Andereck et al., 2005; Látková & Vogt, 2011; Lee et al., 2010; McGehee & Andereck, 2004; Perdue et al., 1990A). Essentially the more personal benefits a resident receives from tourism, the more positive attitudes they possess towards tourism. In one of the few studies to consider resident attitudes towards cannabis tourism (Kang & Lee, 2018), the authors found that personal benefits from tourism was the most important variable when considering resident support for cannabis tourism.

Past research on resident attitudes provides background to understanding resident views on cannabis tourism. Understanding how residents perceive cannabis tourism in their communities has a two-fold impact. First, the study extends research on resident attitudes by investigating whether the variables utilized in resident attitude research can be applied to cannabis tourism. Second, stakeholders can utilize the information to determine specific resident concerns and devise ways to mitigate those concerns and promote cannabis tourism in a way where resident desires are also considered.

Results

Due to the exploratory nature of the study, correlations and cross-tabulations were performed on the data. Numbers of respondents are referenced in the data figures below and vary due to the stage of completion of individual questionnaires.

In considering the respondent's answers on potential related impacts of cannabis tourism, attitudes tended to spread across the agreement scale (Table 3). Many respondents were unsure about the cannabis tourism attitude statements. Just over 50% of respondents tended to agree that cannabis tourism would have some benefits for the county and 51% felt it was a good opportunity. While a majority of respondents (59.2%) did not feel outdoor and recreational tourism would decrease due to cannabis tourism, it is important to note that there was concern among respondents about family-oriented tourism. 50% of respondents tended to agree or strongly agree that some family-oriented travelers would not visit Humboldt County due to cannabis tourism. Nearly 43% of respondents disagreed that the image of their community would be negatively affected by cannabis tourism, however,

Table 2. Characteristics of respondents.

Characteristics	Respondents %
Gender (n=769)	
Female	50.3
Male	49.7
Mean Age (n=762)	47.3 years
Mean Length of Residence (n=794)	19.8 years
Household Income (n=758)	
Less than $25,000	13.3
$25,000 - $49,999	22.2
$50,000 - $74,999	21.6
$75,000 - $99,999	15.0
$100,000 or more	27.8
Education (n=780)	
Less than High School	0.6
High School Graduate	3.8
Technical School Degree	2.1
Some College	24.4
College Degree	44.9
Advanced Degree	24.2

30% of respondents were also unsure if cannabis tourism would positively affect the image of their community.

Pearson correlations were performed between age, length of residency, personal benefit from cannabis tourism, and attitudes toward cannabis tourism.

Among respondents, there was a negative association between age and attitudes towards cannabis tourism. Though the correlations showed overall weaker relationships, the older the participant the less positive their attitudes were towards cannabis tourism (Table 4).

We see similar results when considering length of residency and attitudes (Table 5). Respondents with longer lengths of residency had less positive attitudes towards cannabis tourism, though again the relationships were weak.

We see similar results when considering length of residency and attitudes (Table 5). Respondents with longer lengths of residency had less positive attitudes towards cannabis tourism, though again the relationships were weak.

In terms of personal benefit from cannabis tourism and attitudes towards cannabis tourism, there were positive correlations with moderate to strong relationships (Table 6) suggesting that respondents with higher levels of personal

Table 3. Cannabis tourism attitude items.

Attitude statements	**Strongly disagree (%)**	**Disagree (%)**	**Unsure (%)**	**Agree (%)**	**Strongly agree (%)**
Some family-oriented travelers will not visit due to cannabis tourism	6.7	20.0	23.3	32.5	17.5
Outdoor and recreational tourism will decrease because of cannabis	22.1	37.1	21.3	11.4	8.2
Cannabis tourism benefits Humboldt County	13.1	12.6	23.4	31.0	19.8
Cannabis tourism is a good opportunity	15.7	12.6	20.3	29.1	22.2
The image of my community will be negatively affected by cannabis tourism	16.3	26.6	21.8	17.0	18.4
Out of state visitors will have a negative perception because of cannabis tourism	15.4	28.3	24.3	17.1	14.9
The image of my community will be positively affected by cannabis tourism	19.6	19.8	30.8	19.1	10.8

n=806

Table 4. Pearson correlation between age and attitudes toward cannabis tourism.

Scale	**Age**
Age	-
Some family orientated travelers will not visit Humboldt County due to cannabis tourism*	-.109**
Humboldt County will keep attracting outdoor and recreational visitors	-.154**
Out of state visitors will have a negative perception because of cannabis tourism*	-.163**
The image of my community will be positively affected by cannabis tourism	-.189**
The image of my community will be negatively affected by cannabis tourism*	-.192**
Cannabis tourism benefits Humboldt County	-.198**
Outdoor and recreational tourism will decrease because of cannabis tourism*	-.219**
Cannabis tourism is a good opportunity for Humboldt County	-.226**

*n= 759; * Reverse coded items; ** Correlation is significant at the 0.01 level (2-tailed)*

Table 5. Pearson correlation between length of residency and attitudes toward cannabis tourism.

Scale	**Length of residency**
Length of residency	-
Some family orientated travelers will not visit Humboldt County due to cannabis tourism*	-.148**
Cannabis tourism benefits Humboldt County	-.214**
Outdoor and recreational tourism will decrease because of cannabis tourism*	-.225**
Out of state visitors will have a negative perception because of cannabis tourism*	-.227**
Humboldt County will keep attracting outdoor and recreational visitors	-.227**
The image of my community will be negatively affected by cannabis tourism*	-.234**
The image of my community will be positively affected by cannabis tourism	-.253**
Cannabis tourism is a good opportunity for Humboldt County	-.271**

*n= 790; * = reverse coded items; ** Correlation is significant at the 0.01 level (2-tailed)* benefit from cannabis tourism also had more positive attitudes towards cannabis tourism.

Cross-tabulations were performed to help develop the profile of respondents in terms of perceived knowledge of the tourism industry, personal benefit from tourism, and attitudes towards cannabis tourism. In terms of level of knowledge, participants were asked to describe their level of knowledge by categorizing their perceived level of knowledge from no knowledge to very knowledgeable. Overall data patterns were similar across attitude items. Select results are presented due to space limitations.

Regarding knowledge of the tourism industry (Table 7-8) considering if cannabis tourism was a good opportunity for Humboldt County, of those who indicated they were not at all knowledgeable about tourism, 34.7% were unsure and 34.7% agreed with the statement. As knowledge of tourism increased, respondents were more opinionated with 25% strongly disagreeing and 26.8% strongly agreeing that cannabis tourism is a good opportunity for the county. Considering the statement on positive affect on community image, opinions tend to split once again. Among the very knowledgeable respondents, 33% strongly disagreed and almost 20% strongly agreed that community image would be positively affected by cannabis tourism.

In terms of personal benefit from tourism in general, not cannabis tourism, (Table 9-10) of respondents who indicated no personal benefit from tourism, 25.8% strongly disagreed that cannabis tourism was a good opportunity for Humboldt County, while 28.9% agreed with the statement. As we would expect, respondents who indicated the highest levels of personal benefit from tourism tended to agree (22.7%) or strongly agree (37.8%) that cannabis tourism is

Table 6. Pearson correlation between personal benefit from cannabis tourism and attitudes toward cannabis tourism.

Scale	**PerBen**
Personal benefit from cannabis tourism	-
Some family orientated travelers will not visit Humboldt County due to cannabis tourism*	.357**
Outdoor and recreational tourism will decrease because of cannabis tourism*	.417**
Cannabis tourism benefits Humboldt County	.666**
Cannabis tourism is a good opportunity for Humboldt County	.677**
The image of my community will be negatively affected by cannabis tourism*	.563**
Out of state visitors will have a negative perception because of cannabis tourism*	.519**
Humboldt County will keep attracting outdoor and recreational visitors	.455**
The image of my community will be positively affected by cannabis tourism	.670**

*n= 805; * Reverse coded items; ** Correlation is significant at the 0.01 level (2-tailed)* a good opportunity for Humboldt County. When considering if the image of their community would be positively affected by cannabis tourism, of respondents with no or low levels of personal benefit from tourism, 20.4% disagreed and 31.6% were unsure about the statement. For respondents who indicated higher levels of personal benefit from tourism, 28.6% were also unsure if cannabis tourism would positively impact community image, although we see more of these respondents agreeing (21.1%) or strongly agreeing (20.5%) with the statement.

Discussion

Understanding resident attitudes toward tourism is an important area of study when we consider the potential negative and positive impacts tourism can have on a community. If a community is aware of how residents perceive impacts and the factors influencing attitudes, tourism development can be planned to improve tourism in the community and the quality of life for residents. A large number of studies have been conducted on resident attitudes toward tourism but very few have covered cannabis tourism.

The purpose of the study was to explore if variables used to understand resident attitudes towards tourism would show similar patterns when applied to cannabis tourism in Humboldt County, California. A comparison between the results of past research and the current study are presented below.

Table 7. Cross-Tabulation knowledge of tourism and attitudes toward cannabis tourism. Cannabis tourism is a good opportunity for Humboldt County.

	Not at all knowledgeable (%)	Slightly to moderately knowledgeable (%)	Very knowledgeable (%)
Strongly disagree	12.2	14.3	25.0
Disagree	6.1	12.7	15.2
Unsure	34.7	19.8	17.0
Agree	34.7	31.0	16.1
Strongly agree	12.2	22.2	26.8

Table 8. Cross-Tabulation knowledge of tourism and attitudes toward cannabis tourism. The image of my community will be positively affected by cannabis tourism.

	Not at all knowledgeable (%)	Slightly to moderately knowledgeable (%)	Very knowledgeable (%)
Strongly disagree	16.3	17.4	33.0
Disagree	16.3	20.5	17.9
Unsure	44.9	31.6	19.6
Agree	14.3	21.1	9.8
Strongly agree	8.2	9.5	19.6
n=806			

Age and length of residency. Past research indicates conflicting results in terms of the age of respondents, length of residence, and attitudes towards cannabis tourism (Andereck et al., 2005; Hao et al., 2010; McCool & Martin, 1994; McGehee & Andereck, 2004; Sinclair-Maragh, 2017; Weaver & Lawton, 2001). Regardless, correlations were performed on the data as it was suspected that attitudes towards cannabis tourism would vary with age due to the recent legality of recreational cannabis in California and the changing acceptance of cannabis. In the current study, age was negatively correlated with attitudes towards cannabis tourism. In addition, length of residency was negatively correlated with attitudes towards cannabis tourism. This differs from past research on tourism attitudes, though the relationships were weak. There was some concern from older and more long-term residents that Humboldt County would lose outdoor and recreational tourists because of cannabis tourism. This is important to note as outdoor recreation is one of the biggest draws to Humboldt County.

As noted above, past research on length of residency and attitudes towards tourism has not shown any pattern (Andereck et al., 2005; Hao et al., 2010; McCool & Martin, 1994; Sinclaair-Maragh, 2017). Cannabis tourism may be seen differently than other forms of tourism particularly when considering the age and length of residency of an individual. Due to the long, mostly illegal history of cannabis in Humboldt County, opinions about cannabis tourism vary greatly. Some communities are actively pursuing cannabis tourism while

Table 9. Cross-tabulation personal benefit from tourism and attitudes toward cannabis tourism. Cannabis tourism is a good opportunity for Humboldt County.

	Not at all (%)	Very little to some benefit (%)	Quite a bit to a lot of benefit (%)
Strongly disagree	25.8	14.2	13.5
Disagree	14.4	12.4	11.9
Unsure	16.5	23.1	14.1
Agree	28.9	32.0	22.7
Strongly agree	14.4	18.3	37.8

Table 10. Cross-tabulation personal benefit from tourism and attitudes toward cannabis tourism.The image of my community will be positively affected by cannabis tourism.

	Not at all (%)	Very little to some benefit (%)	Quite a bit to a lot of benefit (%)
Strongly disagree	24.5	19.4	17.3
Disagree	20.4	22.1	12.4
Unsure	31.6	31.5	28.6
Agree	16.3	19.0	21.1
Strongly agree	7.1	8.0	20.5
n=797			

others have made moves to limit it. Cannabis is still federally illegal and many people continue to have a stigma against cannabis.

Level of knowledge of the tourism industry. In past research, no clear patterns have been found between the level of knowledge of tourism and resident attitudes toward tourism (Andereck et al., 2005; Látková & Vogt, 2011) . The current study had similar results when we break the respondents into groups based on their perceived level of knowledge of tourism. In this study, participants who described themselves as very knowledgeable about the tourism industry had more conflicting attitudes regarding cannabis tourism when compared to respondents with none to moderate knowledge of tourism.

Personal benefit from tourism and/or cannabis tourism. In general, past research on resident attitudes towards tourism has shown that if an individual experiences personal benefit from tourism they will likely have more positive attitudes towards tourism (Andereck et al., 2005; Látková & Vogt, 2011; Lee et al., 2010; McGehee & Andereck, 2004; Perdue et al., 1990A). Similar to past studies, personal benefit from cannabis tourism was positively correlated with attitudes towards cannabis tourism. Also, respondents who expressed moderate to heavy benefit from tourism in general had more positive attitudes towards cannabis tourism, as we might expect based on past research.

A majority of respondents agreed with positive statements regarding the opportunity that cannabis tourism presents to the county. However, when we break down respondents based on age, length of residency, level of knowledge and personal benefit from tourism, other data patterns in answers reveal that respondents were unsure about their attitudes towards cannabis tourism and the possible opportunities and challenges cannabis might pose to tourism. It is hoped that stakeholders in Humboldt County will use this baseline study to determine specific resident concerns and devise ways to address concerns and promote cannabis tourism in a way where residents' attitudes are also considered.

Recommendations And Limitations

Currently, there is limited research considering resident attitudes towards cannabis tourism. In order to better guide community and industry responses, managerial implications and areas of future research are discussed below.

The mixed support presented in the current study suggests that some variables may present different results when considering cannabis tourism specifically. This may in turn have managerial implications for destinations pursuing cannabis tourism, like Humboldt County. Uncertainty about cannabis tourism among residents suggests that communities interested in pursuing cannabis tourism should focus on the education of residents. For example, education could center on (1) defining cannabis tourism; (2) the structure of cannabis tourism within a community; (3) the types of cannabis tourists a community will attract; (4) the potential positive and negative impacts of cannabis tourism; (5) the potential negative impacts being mitigated.

Specific to Humboldt County, California, this study was conducted in 2017, after the legalization of recreational cannabis but before legal sales of recreational cannabis started in 2018. Further research might focus on the following questions: What has changed since 2018? Has Humboldt County seen the economic benefits of tax revenues? What challenges have cannabis tourism businesses experienced related to legalization? Another area of consideration is the practical response to resident attitudes. As other US states legalize recreational cannabis and begin to pursue cannabis tourism, it will be important to determine differing attitudes among community members and if government and community actions regarding cannabis tourism accurately represent community views.

In regards to limitations of the current study, a convenience sample was used so the results do not necessarily apply to all residents of the county. As well, the study was restricted to a rural geographic area internationally known for cannabis production. Naturally, this topic generates diverse opinions. Additionally, due to the long history of cannabis in Humboldt County, the opinions of respondents in this study are not necessarily typical of other destinations.

Acknowledgements

Special thanks to the Spring 2017 REC 365 Data Collection Team: Michael Anhorn, Logan Ashdale. Genesea Black-Lanouette, Michael Chavez, Casey Cruikshank, Zachary Dalby, Sky Erbert, Levi Goodeyon, Wes Hewitt, Ian Marting, Jorge Rivera.

References

Andereck, K. L., Valentine, K. M., Knopf, R. C., & Vogt, C. A. (2005). Residents' perceptions of community tourism impacts. *Annals of Tourism Research*, *32*(4), 1056-1076.

Claesgen, D., & Kraft, M. (n.d.). *Workforce Report Humboldt*

County's New Cannabis Landscape (Rep.). Humboldt County Workforce Development Board.

Cui, X., & Ryan, C. (2011). Perceptions of place, modernity and the impacts of tourism–Differences among rural and urban residents of Ankang, China: A likelihood ratio analysis. *Tourism Management*, *32*(3), 604-615.

Economic Effects of Colorado's Legalization of Cannabis 2017. (2016, December 29). Retrieved September 22, 2017, from https://eufloracolorado.com/economic-effects-of-colorados-legalization-of-cannabis-201

Get Ready for Marijuana Legalization, or Get Ready to Lose. (2014, November 22). *Eureka Times-Standard*. Retrieved October 10, 2017, from http://www.times-standard.com/

Giraudo, R. F. (2019). Blazing a Trail: Cannabis Tourism in the United States. In McGettigan, T. (Ed.), *The politics of marijuana: a new paradigm* (pp. 105 - 112). Peter Lang

Gursoy, D., Jurowski, C., & Uysal, M. (2002). Resident attitudes: A structural modeling approach. *Annals of Tourism Research*, *29*(1), 79-105.

Gursoy, D., & Rutherford, D. G. (2004). Host attitudes toward tourism: An improved structural model. *Annals of Tourism Research*, *31*(3), 495-516.

Haley, A. J., Snaith, T., & Miller, G. (2005). The social impacts of tourism a case study of Bath, UK. *Annals of Tourism Research*, *32*(3), 647-668.

Hao, H., Long, P., & Kleckley, J. (2011). Factors predicting homeowners' attitudes toward tourism: A case of a coastal resort community. *Journal of Travel Research*, *50*(6), 627-640.

Houston, W. (2016, October 17). Creating Humboldt County's marijuana tourism market. *Eureka Times-Standard*. Retrieved August 15, 2017, from http://www.times-standard.com/

Jurowski, C., & Gursoy, D. (2004). Distance effects on resident's attitudes toward tourism. *Annals of Tourism Research, 31*(2), 296-312.

Kang, S. K., & Lee, J. (2018). Support of marijuana tourism in Colorado: A residents' perspective using social exchange theory. *Journal of Destination Marketing & Management*, *9*, 310-319.

Kang, S., Miller, J., and O'Leary, J., (2016a) "One year after legalized cannabis: Residents' image, place attachment, and support of marijuana tourism in Colorado". *Tourism Travel and Research Association: Advancing Tourism Research Globally*. 27.

Kang, S., O'Leary, J., & Miller, J. (2016b). From Forbidden Fruit to the Goose That Lays Golden Eggs: Marijuana Tourism in Colorado. *SAGE Open*, *6*(4).

Látková, P., & Vogt, C. A. (2012). Residents' attitudes toward existing and future tourism development in rural communities. *Journal of Travel Research*, *51*(1), 50-67.

Lee, C. K., Kang, S. K., Long, P., & Reisinger, Y. (2010). Residents' perceptions of casino impacts: A comparative study. *Tourism Management*, *31*(2), 189-201.

Light, M. K., Orens, A., Lewandowski, B., & Pickton, T. (2015). *Market size and demand for marijuana in Colorado*. Marijuana Policy Group.

Madrigal, R. (1993). A tale of tourism in two cities. *Annals of Tourism Research*, 20(2), 336-353.

Martin, S. R. (1995). Montanans' attitudes and behavioral intentions toward tourism: Implications for sustainability. *Linking Tourism, the Environment, and Sustainability: Topical Volume of Compiled Papers from a Special Session of the Annual Meeting of the National Recreation and Park Association, 1994: Minneapolis, MN, October 12-14, 1994*, (323) 69.

McCool, S. F., & Martin, S. R. (1994). Community attachment and attitudes toward tourism development. *Journal of Travel Research, 32*(3), 29-34.

McGehee, N. G., & Andereck, K. L. (2004). Factors predicting rural residents' support of tourism. *Journal of Travel Research*, 43(2), 131-140.

Perdue, R. R., Long, P. T., & Allen, L. (1990). Resident support for tourism development. *Annals of Tourism Research*, 17(4), 586-599.

Sharpley, R. (2014). Host perceptions of tourism: A review of the research. *Tourism Management*, *42*, 37-49.

Sinclair-Maragh, G. (2017). Demographic analysis of residents' support for tourism development in Jamaica. *Journal of Destination Marketing & Management*, *6*(1), 5-12.

Taylor, L. L. (2019). Defining Marijuana Tourism. *Journal of Hospitality & Tourism Research*, *43*(3), 438-446.

Travel info for the Redwoods. (n.d.). Humboldt County Visitors Bureau. Retrieved December 30, 2020, from https://www.visitredwoods.com/things-to-do/shopping/

Weaver, D. B., & Lawton, L. J. (2001). Resident perceptions in the urban–rural fringe. *Annals of Tourism Research*, *28*(2), 439-458.

Williams, J., & Lawson, R. (2001). Community issues and resident opinions of tourism. *Annals of Tourism Research*, *28*(2), 269-290.

A Clash of Cultures: The Struggle of Native Americans to Participate in Traditional Ecological Knowledge and Western Science Under California's Marine Life Protection Act

John W. Corbett (Humboldt State University), Ruthie A. Maloney (Humboldt State University)

Keywords: Native Americans, Marine, MLPA, MPA, stakeholder, science-based management

European settlers encountered developed Native American cultures living a nearly idyllic life centered on the bounty of the marine environment and salmon in the rivers of Northwest California. Despite the horrific events of the 19th and 20th centuries, this section of California still supports vibrant, federally recognized, and unrecognized, tribal communities living near the sea. Traditional subsistence harvesting is ongoing for food and ceremonial items, which are bartered with inland Tribes. The Yurok Brush Dance, the Wiyot World Renewal, the Tolowa Dee-Ni Nation Feather Dance, the Pomo Celebration Dance, and other traditional ceremonies, are still practiced by various North Coast (NC) Tribes. The staff of Tribal governments vary greatly. Some have scientists, cultural departments, and lawyers. Significantly for this study, Tribes have participated in federal marine sanctuary planning but continue to face the delegitimization, exclusion, and exploitation of settler colonialism.

The pattern of discounting Indigenous epistemologies and practices is visible everywhere in environmental discourse (Bacon). Of particular concern for this study is the marine planning by the California Department of Fish and Game (CDFG), and their contracted Initiative Science Advisory Team (SAT), systematically excluding Native Americans from providing science input regarding the California Marine Life Protection Act (MLPA). The exclusions have included highly qualified Ph. D. marine scientists and native cultural representatives. It is our assertion that the process completed in 2012 was not fair to all parties, and the actions of the SAT violated the California and U.S. constitution's anti-discrimination provisions, the 1974 Human Research Act, and California open meeting laws. Furthermore, North Coast Study Region SAT models ignored 10,000 successful years of subsistence harvesting and predicted Native take/harvest numbers so high that major marine species would be gone in ancestral territory in a matter of weeks. The failure to allow the participation by native scientists and cultural representatives resulted in science models of lesser quality than those that would have been obtained by the inclusion of Tribal presentations of Traditional Ecological Knowledge (TEK), analytical data developed by Tribes, and an acceptance of Tribal Western science and environmental management.

Attempts to rewrite, or "whitewash," this recent planning history are occurring by claiming Native Americans were not qualified educationally, that Tribal and Western Science could not work together (Olmeta-Schultz), that Native Americans were granted the same participation rights as others, and that the Initiative process provided excellent opportunities for all to be heard. Removing distrust and obtaining support for future reserve planning will require developing a process that all parties agree is fair (Ordonez-Gauger). Native scientist and cultural representatives must be appointed to future science advisory panels.

In 1999, the California Legislature passed the Marine Life Protection Act (MLPA) to establish a statewide reserve system along the entire 1,150-mile California coastline. The MLPA was intended to protect marine resources from overharvest and to establish a scientifically based interconnected system. The MLPA legislation was silent on the intended effects on Native Americans. Private foundation monies and state funding raised 38 million dollars for the "Initiative," a statewide organization whose purpose was to create a master marine plan. A consortium of foundations utilized the Resource Legacy Foundation (RLF) as the sole funder to distribute grant funds. The Initiative agreement was executed in a Memorandum of Understanding (MOU) between the California Resources Agency, CDFG, and RLF. The MOU required the process be "transparent" and "allow for public participation" (California Department of Fish and Game). This agreement did not include sovereign Native American Tribes and the openness provisions were not met by the SAT.

The MLPA legislation divided up the state into four regions. The North Coast study region boundary was the Oregon border to Alder creek Mendocino. Each region had a Blue-Ribbon Task Force (BRTF) to serve as a policy and oversight body, a Stakeholder Group (SG) to draft proposed reserves, and the SAT, which would be tasked with "reviewing and commenting on scientific papers," "addressing scientific questions presented," and reviewing alternative MLPA proposals (California Department of Fish and Game). The results were the creation of a statewide marine reserve system of worldwide significance. California marine reserves, many of which allow no fishing or harvesting of any kind, increased from 1% before the initiative to 16.12% statewide of statewide waters, while north coast marine reserves increased to 13.37%.

In 2010, (NC, BRTF, SG) fishing, environmental, governmental and the vast majority of Native American Tribes came up with a "unified plan" supported by either silence in not objecting or active support (Yaffee). The SAT gave a dismal evaluation of the unified proposal as not meeting science guidelines, for spacing and size of reserves. The unified proposal process started to overcome an impasse with the SAT and dissatisfaction with the planning process. Tribal stakeholders began meeting. Incentives to negotiate were as follows: (a) fishers ("fisher" is a gender neutral term for "fisherman") wanted to participate to limit the number and size of marine reserves, but also had sympathy with Indigenous peoples (Olmeta-Schultz); (b) Tribes' desired to bypass the SAT, preserve gathering rights, and get governmental recognition; and (c) Environmentalists concerned about blocking fisher low reserve proposals (Yaffee) and sympathies with Native peoples (Olmeta-Schultz). According to Poncelet, a facilitator for the NC, the rural nature of the area meant everyone was a neighbor and would have to live with them after the process, which proved to be a strong motive (Yaffee). A distrust of the SAT process also fueled the desire to work together locally (Yaffee). Motivations by each party to reach agreement, as occurred here, are necessary for a stable Marine Life Protection agreement (McCreary). This resulted in the only durable unified plan in the state and the first original master marine plan to designate Tribal ancestral areas.

Key concepts of the unified plan were a reservation of rights to allow for future Tribal challenges and the designation of reserves. Tribal fishing was allowed in some conservation areas, and for the first time, many marine ancestral native territories were recognized on official CDFG maps. However, Tribal fishing requires a license and must conform to state recreational fishing gear types, catch limits, and seasonal closures. Ultimately, important Tribal rights regarding management, co-management, subsistence harvesting, rights to present Traditional Ecological Harvesting, as well as Western science were not settled.

Our goal is to outline where the process went wrong and what is needed to remedy it. Throughout the process, the SAT deliberately excluded Native voices, did not adhere to best science practices of inclusiveness, they authorized and approved their own models, displayed a lack of compliance with public meeting laws and an ignorance of Native perspectives. Level of Protection (LOP) numbers were inflated by catch studies that were not plausible. This egregious disregard for Native scientists and Native perspectives left California with a flawed and inadequate scientific result. Reforms, including anti-discriminatory provisions, must be made. Best Available Science (BAS) guidelines that include standard inclusionary provisions used by federal agencies need to be adopted. Ironically, the CDFG Code Section 33 defines credible science as requiring inclusiveness (Code). Similarly, Fish and Game uses inclusionary BAS for all other regulatory rule making except for the SAT MLPA reserve designation process. A public apology to the Tribes and the establishment of a clear, open, and participatory science process is the surest way to restore trust (Ordonez-Gauger).

From 2005 to 2010, the Yurok Tribe was actively pursuing a federal marine salmon sanctuary with the support of National Oceanic and Atmospheric Association (NOAA) fisheries (Congress, 16 USC Section 1431 Marine Sanctuaries et seq.). There were numerous trips to Washington D.C. to get agreement from NOAA. This started by having NOAA designate the mouth of the Klamath River as a biologically significant area for research to support the creation of a marine sanctuary. The creation of a joint NOAA and Yurok Tribe project to monitor ocean species was agreed to. To carry out these agreements, the federal science research vessel and remote underwater videos were jointly staffed by Native Americans and NOAA fisheries personnel for phase one monitoring. Stakeholders for the federal reserve planning process were contacted and stakeholder planning meetings were held. This planning effort was abruptly stopped in 2010 without notice or apology when the CDFG requested that NOAA stop the program, lest it "interfere" with the MLPA Initiative process. The state ban on Tribal planning has never been lifted. Five years of substantial expenditures spent by the Tribe on this planning process were lost.

A common error in marine planning with Indigenous people is the failure to adjust to cultural differences and differing staff capacities of Tribes (Singleton). The SAT failed to recognize Tribes and their cultural and scientific staffs and, as a result, Tribes were not allowed to present to the SAT. Tribal participation must be an integral component of future marine planning. The lack of participation left intact model conclusions contrary to the peoples' coastal way of life, and model assumptions exaggerated the Native harvest in outlandish ways.

Public interest advocates have been criticized for defining the underlying causes of environmental problems in things such as technical deficiencies, rather than a difference in values (Shellenberger) (Mazur). To implement an advocacy coalition framework, the MPA stakeholders must be studied, and a good understanding of the political context and the values of the parties be ascertained. Christopher Weible contends this was not done by the MPA marine plans. As predicted, this creates suspicion between the parties and the projection of maliciousness on disagreeing parties, as Weible says, "true technical marine science can only occur after recognizing the value conflicts of the parties" (Weible).

Many Tribes live a traditional life, of native traditions, foods, gathering, and use native tongues, yet they must also function in the world of Western culture, analytical science, the English language, and U.S. judicial systems. The mastery of Western ways of science was necessary to protect federally recognized fishing rights created first by a Presidential executive order and then by the 1988 Hoopa Yurok Settlement Act 25 USC 14. These rights must be asserted and defended in federal agency meetings, administrative hearings, and court nearly every year and are primarily based on Western science. To lose the right to participate in the science of these proceedings would create an existential crisis.

Tribes have found that when they supplemented TEK with their own Western analytical and modeling science, they did better in court than solely relying on the Western science of the federal government and water agencies. For decades, many Tribes have supplied science data to the North Coast Air Quality Regional Board, California Public Health, U.S. E.P.A., North Coast Regional Water Quality Control Board, and the Pacific Coast Fisheries Commission. Tribes mistakenly assumed their scientific participation would be welcomed by the SAT. Hillemeier, a Fisheries Director for the Yurok Tribe said in a statement to the SAT: "We really ought to be allies, and we're very distressed, discouraged, and challenged that hasn't been how we have been greeted" (SAT). A science process that excludes tribes and their knowledge is unacceptable.

Clearly any assessment of the SAT science process is dependent on an accurate historical record. Such a record regarding the SAT does not currently exist. This document is intended to contribute to a true history of this process. Past reviews have assumed in published descriptions an open and free process of the SAT: "*The Science Team's process...was open to the public...with ample opportunity for interaction with the public*" (Saarman); "ensuring local stakeholder perspectives…multiple opportunities for public participation existed" (Kirlon); "successfully navigate challenges to public policy science" (Fox); "The Initiative provided numerous opportunities for broad involvement" (Gleason); "...numerous opportunities for participation" (Sauyce). Such an open process did not occur with the SAT, and highly qualified Native American scientists were excluded from participation. Ex ad Hoc rationalizations by the SAT leadership still deny the reality of turning away quality tribal presentations.

The basic structure of the SAT advisory team was an independent science body, free from political interference, in order to provide a marine planning process that was science driven. The MLPA 2855 (c) provides for a Team having one member from Parks and Recreation, CDFG, and the State Water Resources Control Board (SWRCB). The legislature provided for five to seven additional scientists, selected to serve at the pleasure of the CDFG Director. The Director then added ten additional scientific positions until the Team consisted of 21 members. The SAT consisted of 62% academics, 24% were from state agencies, 9.5% private consultants and 4.5% environmentalists and 14 % were women. There were no Afro-Americans, Asians, Latinos, or Native Americans.

The SAT had the common challenge of using science in the context that the final result was mandated by the MLPA. This requires a delicate scientific balance to maintain objectivity within the legislative guidance. The challenge became even greater as some of the scientists had contracts related to the initiative process to make reports that would be used by the SAT. Studies show contacts by SAT members were the highest with environmental groups and the State Government (Weible). The challenge to the science was even greater because many of the scientists working on these reports were also members of the pro MPA advocacy coalition. Of course, a science panel can have both supporters and opponents to the process, but extra efforts need to be made to show fairness to the public. Generally, having members from an advocacy group makes the task of showing fairness very difficult. Turning away citizen concerns by stating the SAT doesn't have conflict of interest appeals suggested great sensitivity. SAT efforts, to ensure public confidence, if any, were enfeebled. The technical marine science can only occur after recognizing the true value conflicts of the participant's (Weible).

Table 1.

MLPA Elite Scientist	Central Coast	North Central Coast	North Coast	South Coast	OST	Chairman	Total
Carr	Central Coast Chairman (1) MPT (1)	North Central Coast Chairman (1) Chairman (1) MPT (1)	North Coast Chairman (1) Chairman (1) MPT (1)	South Coast Chairman (1) MPT (1)	1	1	**12**
Bjorkstedt		North Central Coast Chairman (1)	North Coast Chairman (1) Chairman (1) MPT (1)	South Coast Chairman (1)			**5**
Costello		North Central Coast Chairman (1)	North Coast Chairman (1)	South Coast Chairman (1) MPT (1)	1		**5**
Morgan		North Central Coast Chairman (1) Chairman (1) MPT (1)	North Coast Chairman (1)	South Coast Chairman (1)			**5**
Murray	Central Coast Chairman (1)		North Coast Chairman (1)	South Coast Chairman (1) Chairman (1)	1		**5**
Neilson		MPT (1)	North Coast Chairman (1)		1	1	**4**
Gregoria		North Central Coast Chairman (1) MPT (1)	North Coast Chairman (1)	South Coast Chairman (1)			**4**
Scholz		North Central Coast Chairman (1) MPT (1)	North Coast Chairman (1)	South Coast Chairman (1)			**4**
Gaines	Central Coast Chairman (1) MPT (1)	MPT (1)		South Coast Chairman (1)			**4**

Most of the SAT work was done privately in a committee system with no agenda, minutes, or papers available to the public. Without any representation on the Team, Tribes were dependent on the public meeting process and the right to present written papers. Such presentations never materialized.

In contrast to the makeup of the SAT, the BRTF and SG had Native participation and worked to be inclusive. None of the SAT scientists appeared to be familiar with TEK. The most frequent background favoring appointment to the MLPA SAT was affiliation with Partnership Interdisciplinary Studies of Coastal Ocean (PISCO). According to the PISCO website. " The Partnership for Interdisciplinary Studies of Coastal Oceans is an academic consortium that conducts research to advance understanding of the coastal ocean within the California marine ecosystem and inform management and policy." There was an elite group of scientists who moved from one Regional SAT to the next Regional SAT.

The Team Co-Chair, who was one of the most active in the censorship of the Yurok Tribe, was subject to an ongoing complaint of science fraud by the Yurok Tribe in matters unrelated to the Marine Initiative (Y. T. Corbett). The SAT staff refused two requests for an ethics review, without any hearing or support information being allowed to be submitted, yet the Co-Chair was subsequently convicted in the federal courts for conspiracy to commit science fraud and was sentenced to prison (North District Case). This created a leadership bias against Tribes during the SAT process. It is recommended that the CDFG provide for ethical appeals for the protection of the public. Most agency science panels do provide systems of ethical reviews to ensure fairness.

Some have suggested that adaptive management and the use of after-the-fact monitoring solves all problems. The CDFG has declared reserve monitoring results to be anecdotal and therefore not acceptable to refute model assumptions. According to CDFG, "even historical records of take (i.e., how many mussels were taken from each and every cove, each year, along the whole North Coast) were available to the SAT but it is still uncertain how this may change" (Game, Letter sent to Yurok Tribe from Becky Ota). Since it is considered unprovable that something won't change, Native Americans will never be allowed to challenge the SAT assumption. Past and future Native American presentations have, and continue to be, denied by this reasoning. The rejection of all West coast mussel studies as insufficient erodes the basic concept of data driven science.

The required 3-minute general meeting/public time was confined to issues that did not involve the SAT agenda and policy matters, i.e., not science data (Team, Public Speaking Regulations). It was recommended that longer comments be put in writing, but all written comments were denied to the Tribes by saying no to all requests to present. The SAT lacked any scientists familiar with anthropology and Native American customs. The SAT adopted the bad policy of excluding written and oral presentations by Tribes, which resulted in immediate significant protests and the development of approaches to bypass the SAT as described below. To obtain public support for the Tribes, stakeholder group member Reweti Wiki, a Māori representing Elk Valley Rancheria, circulated a stakeholder petition to the BRTF. The petition advocated for the "aboriginal right to take marine resources for traditional subsistence, cultural, religious, ceremonial, and other customary purposes" (Yaffee). The petition was signed by all but two RSF members, but also by city councils, Tribes, county supervisors, harbor districts, and environmentalists. This petition was backed by the strongest local support there has ever been for a Native American Rights issue (Olmeta-Schultz).

On June 29, 2010, at the Eureka SAT meeting, there were 75 Native protesters representing Hoopa, Talowa, Wiyot, Karuk, Yurok, and various tribal members who marched on the sidewalk outside of the Red Lion where the meeting was held. About 30 demonstrators subsequently entered the SAT meeting room. Protestors demanded to be included on the SAT. As protest leader Frankie Meyers stated, "We would like to ask the SAT to have a representative of the Tribe on the SAT." Dr. Tucker, Ph D. in Chemistry, and member of the Karuk Tribe natural resources department requested peer review papers. Mr. Colegrove of the Hoopa Tribe, Ms. Stevenson (from Laytonville), and others spoke as well (Team, Public Speaking Regulations). Since they were not on the agenda, the SAT cut off the microphone while Susan Burdick (a Yurok elder) was speaking. Mrs. Burdick continued speaking without a microphone and the meeting was adjourned. After consultation, presumably with Sacramento, about how the SAT should respond, the meeting re-opened, and a total of fifteen minutes was granted to thirteen native representatives. Susan Burdick reminded the room of the historic context: "villages being emptied, then the parks come and take over and try to regulate us" (Burdick). This is the classic order of events in settler colonialism, a multi-stepped process moving from expelling Native Americans from the land, to occupying the space, regulating its use, and discounting native epistemologies (Bacon).

On July 21, 2010, in one of the largest demonstrations in Fort Bragg history, over 300 members of over fourteen Tribes marched through the main street on the way to a meeting of the Blue-Ribbon Task Force. The outpouring of support from the Fort Bragg community was amazing,' said Jim Martin, West Coast Director of the Recreational Fishing Alliance" (Bacher). Cars honked in support. Banners were placed on highway overpasses. "Recreational anglers, commercial fishermen, seaweed harvesters, environmentalists, sea urchin divers and seafood industry workers walked side by side with tribal members in a show of solidarity" (Bacher). After hearing the Native American demonstrators, BRTF members pointedly suggested that a Native person be appointed to the SAT.

On November 17, 2010, there was a quiet demonstration at the Eureka SAT meeting consisting of three Ph.Ds., two holders of master's degrees, and Tribal scientists who stated their qualifications and expressed their regret they had not been allowed to present and that they were looking forward to being able to work together in the future with the scientists. No one offered to place them on the agenda or schedule testimony or invite them to present papers. They were met with complete silence. The official SAT minutes state, "they had all showed up to work together in the future" (SAT). This statement erased the substance of the demonstration from the official minutes.

One SAT member quit attending meetings because of the Levels of Protection (LOP) modeling concerns and the treatment of Native Americans. Various CDFG staffers supported the tribe until told to stop or else they would be disciplined. The Chairman of the joint State Senate and Assembly Marine Affairs Committee supported Tribal rights to present testimony and papers to the SAT. While the SAT did not respond, the community demonstrations proved decisive in winning public and political support. The SAT decision to exclude Native Americans converted the independent, supposedly neutral SAT into a highly politicized body deeply involved in settler colonialism and discriminatory race relationships.

It is important to clarify science standards so the public can understand the process. The science standards to be used by the SAT and the MLPA Initiative were never published in their entirety, causing confusion throughout the process. Most public stakeholders and Tribes favored Best Available Science (BAS), as defined by the National Research Council and the 1976 Magnuson Stevens Fishery Conservation and Management Act. This science standard requires that all points of view be considered (Council). While published authors, the majority of the SAT scientists were unacquainted with public regulatory science law, which requires public hearings to receive testimony from all parties. This is to protect the right to due process of law through review by the courts. The SAT applied many BAS principles, but their definition had no provisions for inclusiveness (Harty).

The legal department of CDFG maintained that the word "readily," in best readily available science, referred not to availability of materials but was a rejection of BAS standards and interpreted MLPA specific legislative science standards to be discretionary (Coast). This standard had no requirement to hear from all parties. This legal opinion was adopted by the CDFG (Commission) and provides an institutional structure that supports the opportunity for discrimination. In one instance, a Yurok Tribal presentation that was prepared well before agenda deadlines with copies, accompanied by peer reviewed articles and a flash drive, were turned away under the legal department standard as not meeting the best readily available science standard. The SAT scientists and California Fish and Game legal departments have persisted in their misinterpretations of BAS to this day.

The NG was the last to take up marine planning. The SAT made a series of decisions to support and encourage the participation of fishers. For example, the SAT welcomed fishers at a hearing. In response to fisher Bob Berchale's public comments, Co-chair Dr. Mark Carr states "he raises a very important point that people are finding, preparing, and making data available to the SAT. The time frame is any time you can get it to us before the final evaluations are done. We will absolutely take a look at that...so please don't feel that the window is closing to get information in" (SAT). Yet, for Tribes, there was no invitation to present papers or give testimony, nor were experienced scientists and policy leaders allowed to testify. Seven Ph.D.'s, four master's degree holders, and tribal cultural representatives were turned away. One of the presenters had a Ph.D. in biology, was a Professor emeritus at Humboldt State University, and had conducted marine studies of the local area since the seventies, and many consider him the most knowledgeable scientist of north coast marine communities. Another presenter, who worked on the appointment of a Native American to the SAT, has a law degree from Yale University, was the former Deputy Assistant Secretary for the U.S. Department of Interior as well as the former California Secretary of Natural Resources, was a Central Coast BRTF member, and was on the Dean's Advisory Committee of the Bren School of Environment, Science, and Management. He is considered an outstanding national leader in resource planning. Another presenter was a Ph D. Anthropologist who is a long-term lecturer at San Jose State University with nine publications and was the Yurok Cultural affairs officer of many years, a Yurok Tribe elder, whose testimony was routinely accepted in public forums. Initiative Director, Ken Wiseman, reduced qualified native scientists and culturists to the role of making policy statements to other Initiative bodies (Wiseman, Executive Director of Initiative).

Taped interviews with top Initiative and SAT leaders, conducted by Dr. Olmeta-Shultz, showed that Tribal scientists were viewed as not having a high enough "academic or professional level that typically made it to the SAT" (Olmeta-Schultz). This statement disregards that the SAT itself does not have all Ph.D. scientists and that published marine Ph.D. scientists were turned down. The SAT staff told the Yurok Tribe that "Indian Science had no credibility" (Aireme). And, by e-mail, a Tribal request to get on the agenda was denied because "tribes had no data to present" (Wiseman, Executive Director Marine Life Protection Act Initiative).

From the Yurok Tribe alone, there were over fifteen papers and presentations that were not heard, fifty-two emails sent requesting to be heard, nine phone call requests, two hand-delivered requests, three meeting video tapes, and nine unanswered letters asking what data and modeling science could be presented/or introduced. Requests for peer review articles from Dr. Tucker of the Karuk Tribe were never responded to (Tucker). Clearly, by comparison to others appearing before the SAT, the aforementioned Native Americans were extremely qualified. No other Ph.D.'s were denied the right to present papers during the process. There were more proposed Native American Ph.D.'s to present science to the SAT than there were for the entire state over the MPA five-year process. This is a clear example of settler colonialism and completely negates an often-heard SAT contention that Native Americans were treated like everybody else.

The SAT explained to the Tribes that meetings were private and so there was no right to a public hearing (Wiseman, Executive Director of Initiative). The reasoning was that while the legislature established the SAT as a public body, the SAT had been changed to a new private entity consisting of the old SAT and additional appointed scientists. It was argued no entity existed that could be sued (Gurney vs. California Department of Fish and Game). The private and public body having the same name was apparently a coincidence. CDFG Legal counsel contended the matter was of a first legal impression thereby requiring a court decision. No matter how soft the voice, Native Americans were labeled as "obstructionists" with a "reputation" because they expressed the view that meetings were public (Olmeta-Schultz). There were many Superior Court rulings on this issue, and they all found the Native Americans were right that SAT meetings were public. As part of an appeal of a trial court ruling The California Appellate court opinion stated… "our conclusion that the Task Force is not a private entity or non-governmental body" (Coastside Fishing Club v. California Fish and Game Commission).

There were five principal reasons for the poor public meeting compliance of the SAT: (1) There was a lack of training and knowledge of public meeting laws; (2) there was inadequate legal advice and oversight over meetings; (3) support staff were untrained in the public meeting process; (4) the Initiative process was extremely complex, and understandability suffered; (5) there was a lack of commitment by the SAT leadership to comply with the public meeting laws. Public meetings laws were considered a "Barnum and Bailey" circus by SAT chairs (SAT).

The Attorney General's *Open Meeting Law Manual* for California was distributed to each NG Team member. The manual was unsuccessful in guiding SAT behavior. The SAT consistently had late notices and filings and was unable to make multiple thick agenda packets available to the public. The SAT required everything be done electronically when the law required that hard copies be provided to those that requested them. Every single meeting had packet changes less than the ten-day Open Meeting Act requirements. Seventy two percent of agenda changes occurred only 48 hours ahead of the meeting and many were on the day of the meeting. Large packets were many times not available to the public. A sportfishing representative, Mr. Greenberg, stated to the SAT, "everything you have been discussing on this document was not available publicly minutes literally…if they can even find it" (SAT). There were hundreds of violations of the open meeting laws in a mere 11 meetings.

Additionally, the SAT decided to independently author new assumptions for the Levels of Protection (LOP) model. The standard practice of using peer reviewed publications avoids the problem of not finishing the model on time. The SAT ran out of time and did not finish the LOP

Table 2.

Violation of Agenda Laws **Bagley Keene Open Meeting Law Illegal (less than 10 days Agenda Revisions)**	**Number**	**Percent**
10 days' notice requirement	0	0%
Emergency finding necessary to legally shorten time	0	0%
Day of meeting	2	18%
One day before	4	36%
Two days before	4	36%
Three days before	0	0%
Four days before	1	9%

**Over 90 % of the notice revisions were for two days or less. *Meeting packets for revisions were commonly not available until the day of the meeting and often were not available for the public.*

model until the very last meeting. The model had no data, published protocols, or complete model assumptions.

The next section will first cover attempts to submit Traditional Ecological Knowledge (TEK) and the following will cover Western science. TEK is defined as a "cumulative body of knowledge, practices, and beliefs, evolving by adaptive processes and handed down through generations by cultural transmission, about the relationship of living beings (including humans) with one another and with their environment" (Ramos). The choice between science and TEK is not either or. Western science is useful to supplement TEK. For instance, science can help in many ways: to ensure that laws and protocols designed to protect Tribes be enforced, such as the Human Research Act of 1974; for situations where the natural environment has been completely changed such as pesticide and nutrient pollution in the Upper Klamath River; to provide an alternative to other Western models that are clearly wrong and adverse to Native American interests; and if the sovereign Tribes wants to make such presentations. However, Western science is no substitute for TEK.

Direct presentations for subsistence harvesting and traditional knowledge were universally turned down by the SAT without any context for local customs and harvesting and, consequently, marine planning for the north coast region suffered. By rejecting Tribally sourced harvesting data available to the Team, subsistence harvesting vanished from consideration in the SAT process. Becky Ota of the CDFG stated to the SAT that what was needed was an anthropological report. An anthropological report is a term often used by Western scientists referring to Tribal practices, and it often implies they are no longer practiced. TEK was turned down because "Indian science" was not credible. This is ironic as the SAT determined that there was insufficient data on Native American harvesting. In response, the Yurok Tribe met with SAT staff to try to determine who could be credible. After turning down a Tribal Ph. D the conversations centered on hiring an outside anthropologist Dr. Jeanine Pieffer. It was the Tribe's understanding that the report would be given in writing to the SAT ahead of time and then presented.

On July 28, 2010, the SAT had an agenda item titled "Review and Discussion of SAT Study Conclusions to Science Questions." Yet, the SAT agenda questions were not available before the meeting so there was no way to know what was going to be discussed. The SAT meeting did answer questions covering important Native American issues, but the notice was so vague and late not a single Native American was in attendance. After the fact, the Yurok Tribe learned the scheduled presentation of the Tribal TEK anthropological report was cancelled, since it was already covered in SAT answers to questions.

E-mail correspondence between John W. Corbett (Attorney for Yurok Tribe) and Sate' Aireme (Principal investigator SAT), describe the lack of Tribal satisfaction with the process:

"The Yurok Tribe is also puzzled that the format which we proposed earlier of presenting a Tribal paper to the SAT as a regular agenda item changed, without notice, to no Tribal input...and the process was going forward without the presentation of our paper." ... In summary, the Yurok Tribe and Native peoples may have lost our opportunity to have informed the SAT with a written report from a qualified consultant. The Yurok Tribe will have spent $6,000 to $10,000 to have a report prepared that can't be submitted to the Science Panel and the Tribe is still not on the agenda" (Y. T. Corbett).

The SAT decided to conduct their own study of TEK and subsistence harvesting. The 1974 Human Research Act requires behavioral research permits to protect the rights of human subjects. There seemed to be no awareness of this requirement to get a permit and the survey proceeded without proper authorization starting on March 3, 2010. The gathering of information cutoff date was July 27, 2010. The data was gathered, and tentative conclusions were being shared. The survey forms had no risk disclosures that this was part of a regulatory program that might criminalize existing subsistence harvesting by putting them into a no-take reserve. In a no-take reserve, no harvesting of any kind or species is allowed and violators can be criminally charged.

Internal Review Board (IRB) rules state the following regarding human research: *"The protocol must be reviewed and approved by the UCSB Board before the research begins. Failure to comply with these rules may have serious consequences, including the suspension or termination of research, allegations of research misconduct, and personal civil and criminal liability.* **PLEASE NOTE THERE ARE NO PROVISIONS FOR RETROACTIVE APPROVAL OF RESEARCH PROTOCOLS"** (Cruz). There are two types of I.R.B. permits: One comes after the full board hearing and there is another option for a conditional permit, called exemption, because a full I.R.B. hearing is not required. Conditional exemptions from a full I.R.B. hearing board revies are usually granted for minor permits, limited risk, and uncontroversial projects. All results from survey forms collected between March 3, 2010, until August 9, 2010, before the exemption permit was granted, are void. Given the absolute prohibition, the Yurok Tribe requested that any new information gathered not be comingled with data before a permit is issued. The SAT ignored this advice, and by mixing the data, it is all tainted and unusable under I.R.B. regulations.

Exemption permit conditions required the I.R.B. Human Subjects Committee be immediately notified if there is "adverse reaction...distress regarding the subject matter or procedures" (Cruz). The numerous protests and demonstrations and legislative criticisms make it hard to imagine a stronger record of an adverse reaction. The required report of controversy was never filed by Satie Airame' as required by permit conditions. The whole purpose of the I.R.B. process is to independently review research projects that can affect the rights of subjects. The 1974 Human Research statute is an essential component of protecting Native cultural rights. No such adherence occurred. There is little doubt that the late date, failure to disclose risk, and the absence of disclosing controversy, resulted in serious violations of the spirit, letter, and substance of the Human Research Act of 1974.

Western science has been defined as the systematic study of the structure and behavior of the physical and natural world through observation and experiment (Google Dictionary). In natural resources, this often takes the form of predictive models. The SAT ignored MLPA (Section 2858) peer review requirements and changed the LOP assumption to the following take/harvest assumption in the model it was writing: "Any extractive activity can occur locally to the maximum amount allowed by federal and state law" (SAT). No data or peer reviewed research of any kind was introduced to support the unique SAT take assumption. The following chart immediately shows what is wrong with LOP harvest numbers in estimating the NA harvest for mussels.

Table 3. CDFG license statistics. Recreational Marine Regulations 2010 (Game, Marine Sport License Statistics).

License Analysis Northern California	**Number**
Resident Fishing License	1,112,783
Non-Resident Annual	9,942
Reduced Fee Veterans License	11,244
Subtotal	1,133,969
1,133,969 x 365 day =	486,898,685
Minus purchase date	135,019,587
	351,879,098

A literal interpretation of the assumption would result in even higher numbers. There are two no-license days a

Table 4. CDFG license statistics. Recreational Marine Regulations 2010 (Game, Marine Sport License Statistics).

License Analysis Northern California	**Number**
Lifetime Fishing 11,639 x 365	4,248,235
1 Day Sport Fishing 529,129 licenses =	529,129
2 Day Sport Fishing 122,493 x 2 days =	244,986
10 Day Non-Resident License 14,081 x 10 =	140,810
Annual Fishing Opportunities	357,671,088
Daily Total	988,140
Hourly Total	41,172

year open to all California citizens for marine fishing that were not counted. Below the age of sixteen no license is needed for marine harvesting. The SAT and CDFG have refused to answer questions regarding the inclusion of these higher numbers. In addition to high license numbers, other contributing factors are long seasons and high daily take numbers for species.

All data is from CDFG marine sport fishing regulations. The vast majority of seasons are 365 days, which requires multiplying 365 by somewhere around a million licenses. Any species with a 365-day season and a daily harvest take of five or more has a harvest in the billions. A review of CDFG historical data usually shows annual catches of far less than one million of a species and there are no billion catch recordings ever. The scale differences of the numbers have existing data contradicting the LOP take projections.

Species are concentrated in high daily allowable catch categories. The number 51 on the chart is used for unlimited harvest, which is one more than the highest permitted harvest, in order for there to be a way to graph it. Of course, the infinite harvest results in immediate extinction. In summary, the high number of nearly a million licensed fishers, long seasons, and high catches contribute to such excessively high numbers for mussel harvest that they are not plausible.

The year-round season for mussels has a limit of 10 pounds per day. This creates a projected annual mussel harvest of 3.577 billion pounds. The Yurok Tribe has an estimated 5,700 members, which results in every man, women and child harvesting for personal consumption 627,493 pounds of mussels a year or 1,993 pounds of mussels per day. Hans Voight, a scientist contracted by the Yurok Tribe to present on mussels, worked with SAT principal investigator Sate Aireme' who provided peer reviewed studies to be used for his mussel report. When his study was completed,

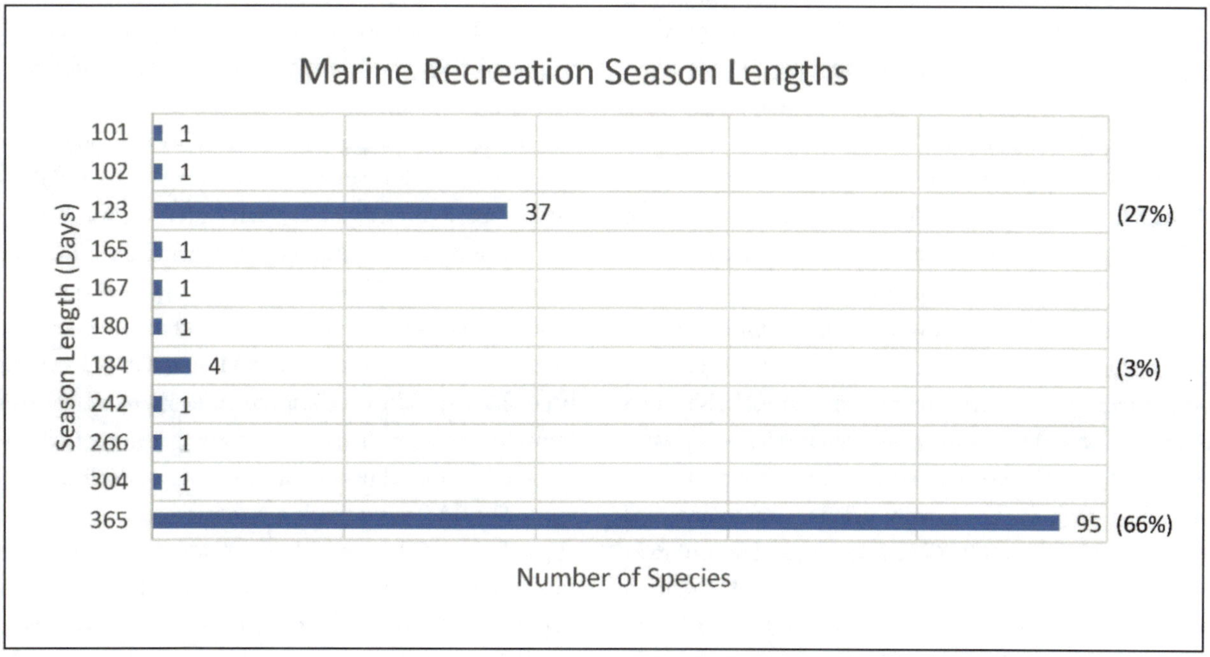

Figure 1.

and his fees paid by the Tribe, the SAT suddenly determined there would be no submittal or agenda presentation. No record could be found of SAT calculations using the LOP assumption model as they were never developed. A complete list of model assumptions is required as a prerequisite for many scientific or legal reviews for the validity of a model formula. Tribes are concerned that these faulty take license numbers project native subsistence harvesting that would result in widespread localized species extinction. The model results are anathema to the reality of subsistence harvesting, are inconsistent with all known studies, and are not a reasonable scientific conclusion.

The SAT failed to develop models that showed the entire population of a species against which the plausibility of the recreation harvest assumptions could be measured. The Yurok wanted the adoption of the peer reviewed Klamath Harvest Ocean Monitoring model (KHOM) figures for total abundance of Chinook salmon in the Klamath Management Zone (KMZ). Any SAT LOP assumption for the subcategory of recreational harvest resulting in more than 50 times the number of the total population should be rejected, due to being so inaccurate it is not of value to marine planning. Tribal calculations showed that the assumption did not pass the test. Consideration of using the KHOM sub-model for predicting recreational harvest of KMZ Chinook was rejected, even though the model follows best practices including peer review and is checked each year for accuracy. As we have argued, it is clear is that the SAT has not established credible harvest projections and has been recommending policy based on these inadequate estimates. There is no doubt their models, which lacked peer review, would have been better with Tribal input.

The Yurok tribe was also concerned about the salmon by-catch numbers. Generally, by-catch calculations are a multi-step process, and it is desirable to use data from the same region. The SAT used a simplified process based upon inadequate sample size in years often as low as four years. All models included the year 2006. CDFG and NOAA have independently found the 2006-year overstated rock fish catch and is not reliable. The central coast data was inappropriately extrapolated to the north coast region without a comparative habitat analysis. The heavily relied upon Commercial Passenger Fishing Vessel (CPFV) data source does not exist as a viable industry in Del Norte County. Specific salmon patterns relating to their river migrations in the KMZ were ignored. The Yurok Tribe desired to present a more complete analysis, specific to the NC region, that rejects the applicability of the North Central study. A recent multiyear Census and Behavioral Survey conducted for False Klamath Cove, shows a projected annual onsite and offsite visitor count of around 250,000 versus the SAT model projection of 358 million fishers. All proposed science-based presentations were rejected. SAT member Craig Strong stated at the January 13, 2011, SAT meeting that, "...the assumption of the maximum allowable take on the North Coast is simply not real and so it renders the whole structure subject to question." When informed that this was the last SAT meeting (1-13-2011), he voted to approve the model anyway. SAT Co-chairman Dr. Eric Bjorkstedt stated: "I think concern was not only reserved to the public with the LOP model. It has been problematic even for the LOP working group because of difficulties (referring to high harvest numbers)" (Bjorkstedt). This is a clear case of the absence of Native American inclusion resulting in an incompetent result.

Conclusion

Before concluding, it should be noted that much of the Initiative and SAT science was of the highest quality. However, as we have argued throughout this article, not all of it was. Marine planning along the California coast came late to this process and on the North Coast the struggle was for Indigenous peoples to participate in the science. The case for TEK could not have been stronger. The SAT, with no members trained in anthropology, rejected all forms of traditional data, subsistence practices, and two TEK reports by qualified Ph.D. anthropologists and Tribal culturists, and then embarked on a survey to substitute their wisdom for that of Indigenous peoples. This effort was not in conformance with the Human Research Act of 1974, violated conditional permit terms, and ended with compromised data.

The SAT self-authored, Western science LOP model fared little better. It was not even completed until the last meeting. No science was ever introduced to support the LOP take assumptions. The SAT and CDFG have never been able to produce a comprehensive list of model assumptions. Such assumptions are an essential prerequisite before a scientific or legal review can even be conducted.

The SAT never made a public calculation of the model predictions. The model harvest assumptions were so high that Native American harvesting was eliminated as being irrelevant. The purpose of the take model assumption was to be able to select where a marine reserve was needed. The SAT assumption of take is so large that every inch of the coast qualifies to establish a reserve. Consequently, the take assumption is of no value in identifying particular reserve sites. The SAT violated the U.S. and State constitutional provisions against discrimination by excluding Native Americans. Given the evidence provided by this article, it is clear that Native American presentations would have greatly improved the model and provide realistic projections.

To an extraordinary degree, the SAT resisted efforts of qualified Native representatives to participate. At the end of an expensive 38 million statewide public participatory effort, the SAT failed to provide a fair system that could build trust and support for the MLPA among Native peoples (Ordonez-Gauger). The dropping of inclusionary provisions from the National Academy of Science BAS definitions provided the institutional opportunity to discriminate. Inclusion needs to be restored as a fundamental BAS principle by CDFG. Referring to the non-use of SAT criticisms of the Unified proposal a key factor was that "The LOP evaluations seemed tainted by the SAT assumptions" (Yaffee). The CDFG Commission needs to recognize and apologize for systematic exclusions of Native Americans. The inadequacies of the LOP model need to be acknowledged along with the clear fact that input from Native people would have improved the science. Indigenous peoples have every legal, scientific, and moral right to participate in the science and management of their homelands.

The behavior is a clear example of settler colonialism and the failure to provide a system of sufficient fairness to build trust. The systematic exclusion of an entire ethnic group and their representatives, coupled with wildly inaccurate harvest and other assumptions, taints the LOP science work of the North Group SAT.

Acknowledgements

Thanks to my wife, daughter, and son. Thanks to Dr. Laurie Richmond of Humboldt State University, building citizen, fishing and Native American participation in Natural Resources, Catherine Khulman former Executive Director of the Ocean Protection Council, Dr. Snezanna Levic, University of London for constant encouragement, Troy Fletcher, The Yurok Tribe, deceased former Executive Director of the Yurok Tribe, scientists Mike Belchik and S Dave Hilliemier, North Coast Surf Fish representative Dennis Mayo, Humboldt Fish Marketing Association representative Vivian Helliwell, Jacque Hostler Fish and Game Commissioner and Executive Director of the Trinidad Rancheria, Blue Ribbon Task Force members Meg Caldwell, juris Doctor currently deputy Director Packard Foundation Ocean Conservation and Science, Roberta Cordera Chumash, Indigenous leadership, Mediator and Jimmy Smith former Humboldt County Supervisor (deceased).

Waklow (Thank You) to my dad, mother, sister, daughter, son and my grandkids. It's honor to advocate for the protection of Indigenous religious, cultural, and environmental rights for future generations. North Coast Native Protectors invite you to join in our work and projects contact Ruthie A. Maloney (707) 502-9155 or northcoastnativeprotectors@gmail.com.

References

1431, United States Code. "Magnuson Stevens Fishery Conservation and Management Act." 1976.

Agency, U.S. Environmental Protection. "Estimated Fish Consumption rates Report." Government. 2010. https://www.epa.gov/fish-tech/estimated-fish-consumption-rates-reports.

Aireme, Sate'. *E-mail to John W. Corbett Attorney for Yurok Tribe.* Klamath, CA, 08 July 2010. E-mail correspondence.

Bacher, Dan. "Tribes and Allies Take Control of Fort Bragg MLPA meeting." *The Fish Sniffer*. Fort Bragg, CA, 24 July 2010.

Bacon, J.M. "Settler colonialism as eco-social structure and the production of colonial ecological violence." *Environmental Sociology* (2019): 59-69.

Bjorkstedt, Eric Dr. "SAT Co-Chairman." *Conversation with John W. Corbett.* Eureka, CA, 13 February 2011.

Burdick, Susan. "Yurok Tribal Member." Eureka. CA, n.d.

California Department of Fish and Game, Resources Agency and Resource Legacy foundation. "Memorandum of Understanding." *MOU.* Sacramento, CA: State of California, 2008.

California Fish and Game Code 2850-2863. "Annual CA Sport Fishery Regulations." Sacramento, CA: 2010, n.d.

—. *Marine Life Protection Act.* Sacramento, CA, 1999.

Carden, Kristin. "Bridging the divide: The role of science in species conservation law." *Harvard Environmental Law Review, Vol. 30* (2006): 186-258.

Chesbro, Wesley. *Chairman Joint Senate Assembly for Marine Affairs.* Letter to SAT. Eureka, CA: Joint Senate Assembly for Marine affairs, July 2, 2010.

—. "Joint Senate and Assembly Committee of Marine and

Aquatic Mandated Legislative Hearing." *Public Meeting*. Eureka, CA, 22 01 2011.

Coast, Marine Life Protection Act North. *Final Environmental Impact Report (FEIR)*. Response to comment 3, Section 3, pg 7-8. Sacramento, CA, 2012, May 03.

Coastside Fishing Club v. California Fish and Game Commission. No. 215 CAl. APP. 4th 397. California Appellate Court. 13 April 2013.

Code, California Fish and Game. "Define Creditable Science." *Section 33*. Sacramento, CA: State of California, n.d.

Commission, California Fish and Game. "North Central Master Marine Plan." *North Group Master Plan*. n.d.

Congress, United States. "16 USC Section 1431 Marine Sanctuaries et seq." *Federal Marine Sanctuary Act*. Washington, DC, 1972.

—. "1974 Human Research Act." *Legislation*. Washington, DC: United States Congress, 1974.

Corbett, John W. *Attorney at Law* 1973.

Corbett, Yurok Tribe. *Yurok Tribe Marine Management Plan*. Managment Plan. Klamath, CAaaaa: Yurok Tribe, 2012.

Council, National Research. *Best Available Science definition*. Washington, DC: National Academy of Sciences, 2004.

Cruz, UC Santa. "https://officeofresearch.ucsc.edu/compliance/services/irb.html." *Delegated authority to administer 45 CFR 46 permits under the 1974 National Research Act usingally referred to as the 1974 Human Research Act*. 2010: The board issued an MSI-AI-SA-007-2N to signed Investigator Satie' Airame, 9 August 2010.

Finance, Department of. *https://esd.dof.ca.gov/dofpublic/viewBcp.html*. 2010. January 2020.

Foundation, National Science. 2004.

—. *Improving the use of the "Best Scientific Information Available" standard in fisheries management*. Washington, DC: National Research Council, 2004.

Fox, Evan., Melissa Miller-Henson., Ugoretz, John., Weber, Mike., Gleason, Mike., Kirlin, John., Caldwell, Meg., and Mastrup, Sonke. "Enabling Conditions to support marine protected area network planning: California's Marine Life Protection Act Initiative as a case study." *Ocean and Coastal Management vol. 74* (2013 March): 14-23.

Game, California Department of Fish and. "Letter sent to Yurok Tribe from Becky Ota ." Sacramento, CA, 12 April 2012.

—. "Marine Sport License Statistics." *Recreational Marine Regulations*. 2010.

—. *Mussel Power/New Zealand Geographic*. n.d. www.nzgeo.com.

Gleason, et al. "Designing a network of marine protected areas in California: achievements, cost, lessons learned and challenges ahead." *Ocean and Coastal Management, Vol. 74* (2013): 90-101.

Google. *Google Dictionary*. 28 June 2021.

Gurney vs. California Department of Fish and Game. No. SCUK CVG 105448. Resources Agency Mendocino Superior Court. n.d. Declaration of Deputy Attorney Genral Counsel.

Harty, Michael J, Dewitt, John. "Report on Lessons Learned from the Marine Life Protection Act Initiative: North Central Coast Study." 2008 October 31.

Jaffee. (n.d.).

Kirlon, et al. "Supporting Implementation of legislation establishing a Statewide work of Marine Protected Areas." *Ocean and Coastal Management, Volume 74* (2013): 3-13.

Legislature, California State. "Joint Committee on Fisheries and Aquaculture." Sacramento, CA, 21 January 2011.

Malloy, Katherine M. "California's Marine Life Protection Act: A Review of the Marine Protected Areas Proposal Process for the North Coast Region." *Southwestern Environmental Law Journal* (2008).

Marine Life Protection Act, North Coast. *Final Environmental Impact Report (FEIR)*. Response. Sacramento, CA: Adopted by the Fish and Game Commission, 2012 May 03.

Mazur, Allen. *The dynamics of technical controversy*. Washington D.C. : Washington D.C. Communication Press, 1981.

McCreary, Scott, Grifman, Philis and Cowart, Meredith. "Case Analyses Creating Stable Agreement in Marine Policy: Learning From The California South Coast Marine Life Protection Act Initiative." *Negotiation Journal* (2016): 23-48.

Miller-Hensen, M. "Joint Senate & Committee of Marine and Aquatic Mandated Legislative Public Hearing." *Marine Life Protection Act Initiative*. Eureka, CA, 21 January 2011.

Minsuk. 2015.

Minsuk, Jun. "Governing Marine Protected Areas (MPAs) in California: Analysis of the MLPA Implementation Process." *Ph. D. Thesis*. London: University College London, 2013.

North District Case. No. 13-CR-00683. U.S. District Court Northern District . n.d.

Olmeta-Schultz, Ph. D. "The North California Coast Marine Protected Area Network: Study of Commercial Fisherman Perceptions of MPA effort." *The Planning Process and Regional Stakeholder Group Collective Action*. UCSC, December 2018.

Ordonez-Gauger, L., Richmond, L., Hackett, J.,. "It's a Trust Thing: Assessing Fisherman's Perception of the California North Group Marine Protected Area Network." *Ocean and Coastal Management* (2018 May): 14-153.

PISCO. "Partnership for Interdisciplinary Studies of Coastal Ocean." 1999.

Ramos, Seafha. "Hikelonah Ue Meggetohl In An Ever Changing World." Cooper-Lara, Kishan & Lara, Walt J. Sr. *Ka'm-t'em A Journey Toward Healing*. Temecula, CA: Great Oak Press, 2019. 85-93.

Saarman, Emily et al. "The Role of Science in supporting Marine Protected Area Network Planning & Design California." *Ocean and Coastal Management Vol 74* (2013): 45-56 pg 47 last paragraph.

(SAT), Science Advisory Team Meeting. "MPA Public Meeting Red Lion." *Meeting Transcripts prepared by an independent certified reporter*. Eureka, CA, 28 July 2010.

Sauyce, K., Shuman, C., Connor, D., Reisewitz, A., Pope, E., Miller-Hensen, M., Poncelet, E., None, D., Owens, B. "Beyond Traditional Stakeholder Engagement: Public Participation rates in California's statewide marine protected area Planning Review." *Ocean and Coastal Management Vol. 74* (2013): 57-66. https://doi.org/10.1016/j.oceacoman. 2012.06.012.

Shelbenberg. 2004.

Shellenberger, Michael and Nordhaus, Ted. "www.thebreakthrough.org/images/Death of Environmentalism.pdf." 04 April 2006. *The Death of Environmentalism: Global warming politics in post environmental world.* 2021.

Singleton, Sara. "Native People and Planning for Marine Protection Areas: How "stakeholder" Process fail to address Conflicts in the complete Real World Environment." *Ocean and Coastal Management Vol.37 (5)* (2008): 425-440.

Staff, California Fish and Game. *Marine Life Protection Act*. Sacramento, July 2010.

Team, Science Advisory. "Public Speaking Regulations." Sacramento: State of California, n.d.

—. *SAT taped meeting video reaffirmed at the July 28, 2010 SAT meeting*. Eureka, CA, 11 February 2010. Video.

Tucker. n.d.

United States Code 1431 et seq and 33. *The Marine Protection, Research and Sanctuaries Act*. Washington, DC, 1972.

Weible, Christopher M. "An Advocacy Coalition Framework Approach to Stakeholder Analysis: Understanding the Political Context of California Marine Protected Area Policy." *Journal of Public Administration Research and Theory* (2006): 95-117.

Wiseman, Ken. "Executive Director Marine Life Protection Act Initiative." *SAT meeting video*. Eureka, CA, 28 July 2010.

—. "Executive Director of Initiative." *e-mail to megan rocha Yurok Tribe from Ken Wisemen*. Sacramento, CA, 19 June 2010.

Yaffee, Steven. *Beyond Polarization Public Process and the Unlikely story of California Marine Protected Areas*. Washington: Island Press, 2020.

Ray Cesaretti: Chosin Reservoir Korean War

Michael Pazeian (Humboldt State University)
(with assitance from Marilyn Cesaretti)

Figure 1.

I was born 10/27/29, soon I will be 90 years old. I was living in Wildwood/Rio Del in northern California at the time of Pearl harbor. I didn't know much. But I got my BB gun, put up some tin cans and shot at them. Each time I hit a can I said that's for you—you Japanese. That is the way we felt. It was a sneak attack. We were somewhat isolated up here, so I didn't keep up with World War II. There were coast patrols and an occasional black out. And there was some rationing of food.

I graduated from high school in 1947 at the age of 17. I went to work for Pacific Lumber Company. But there was a deal that if you went into the reserves for 10 years you would do your service. WWII just finished, so I thought I would be safe to join the reserves. I joined the Marine Reserves. It was 1949. I did my basic training at Camp Pendleton.

The rifle, drilling, bayonet practice, I remember—and the marching. It was springtime and the rattlesnakes had just come out. One I saw, its head was bigger than my fist. Our corpsman got bite by this snake. He survived; he had the anti-toxin. He did get very sick. I became a browning automatic rifleman. I also shot a bazooka. I went through demolition school, explosives. It got hot there before we finished.

I went back to school at Humboldt State. In 1950 the Korean War started, and I was called in. It was September. I went to Treasure Island and waited for our company to become full. We played sports while we waited. It took a week for all the guys to show up. We took a bus to Camp Pendleton. One of the guys was a buddy from Eureka. Doug Curtis would later become a police officer in Eureka. Another guy was from Scotia, Garth Jaehing was also with me in the same unit. We all trained at Pendleton. More training. We found out how to do without sleep and food. And we learned to ration our water. We did a lot of climbing. There was one hill we called Nelle's tit. We had to carry a heavy machine gun on our back and duck walk up the hill. They were going to get our baby fat off and get us in shape in a hurry. We were there two weeks.

We were on a bus to San Diego. On board ship, a troop ship, a liberty ship. Lots of soldiers on it. If you wanted to eat twice in one day you had to stay in the chow line. Bad food. We went to Japan. Most of us didn't go down to the sleeping area. Many were seasick. The bunks were 3 high and the guy on the bottom got all the vomit. The bunks were very closely stacked, you could not roll over. I was on the top bunk. Garth was a radio repairman. There were other ships with us, a destroyer and another ship. At Otsu, Japan there was a band waiting for us, playing as we got off the ship. They played, "If we knew you were coming, we would have baked a cake," which was a popular song by the Fontane Sisters at the time.

We got time to go into the town and drink and visit with girls. Most of us went to the "Slop Shoot," the local beer hall. We were there maybe two weeks then on a ship to Korea. We did physical training and fired our weapons. I fired my browning automatic. It was a 30-calibur machine gun. I had a guy with me as my ammo carrier. Early in combat, I wanted to kill him because he dumped all the extra magazines of ammo. Instead, he was carrying cans of food. I was really mad, the top sergeant stopped me from doing something to him and had him replaced. We arrived at Wonsan, North Korea on the northeast coast.

Our company left Wonsan and walked to Hamhung, approximately 100 miles. We arrived just before dark and found a place to bed down. I was part of the 1st Battalion, 7th Marine Division, Charlie Company. We carried everything we needed on our backs. I had a parka for the cold weather. I spent my 21st birthday, Oct. 27th, in Hamhung. British and Turkish allies were with us. We were there a short time. We were pushing the enemy north. The North Koreans were retreating towards their northern border, the Yalu River. Many North Koreans are surrendering without firing a shot. It was difficult to guard them all.

We pushed north for about a week. As we moved north, resistance got stronger. At times we were firing our weapons, at times in a heavy fire fight. We knew the Chinese were gathering ahead of us. (The Chinese entered the war in late October 1950. Major General Oliver P. Smith, commander of the US 1st Marine Division slowed our northern movement because of the possibility of many Chinese in front of the Americans.) Some of the bodies and prisoners we took in were Chinese. We moved through rock canyons northward. We came out on the plain and see the reservoir. It is very cold—well below zero. In the morning, I would get out of my sleeping bag, and my boots were frozen on my feet. Everyone was suffering from frostbite. I still suffer from it today.

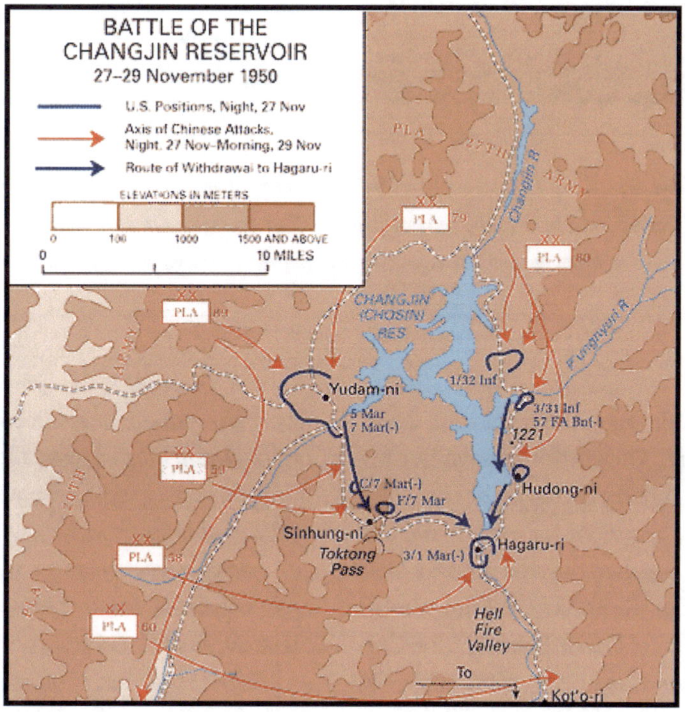

Figure 2.

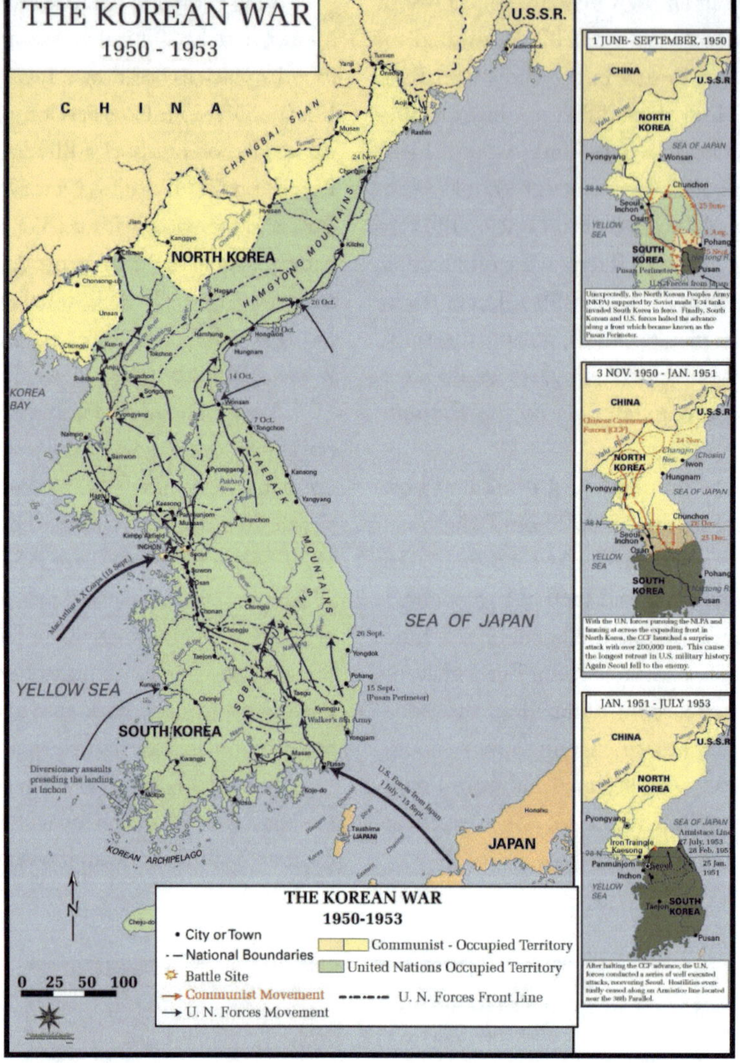

Figure 3. *On the night of November 27, the Chinese attacked the American positions all around the Chosin Reservoir. The 7th Marines were dug in on the ridges surrounding Yudam-ni. The Marines were able to hold their positions. On November 30th, the Division Commanders ordered to break out from Yudam-ni to Hagaru-ri, a strategic withdrawal.*

It is late November 1950. We were surrounded. Charlie Company was surrounded. They broke through our perimeter. I was shooting a lot. A bullet hit me in the chest. But it went through my clothes and bruised my breastbone, and that was all it did. The cold weather, the cold air may have slowed the bullet. I was part of the perimeter. They wanted our food and clothes. There was lots of fighting. The Chinese lost a lot of people. I kept firing my weapon. They did get to our supplies. Most of their attacks were at night. Night after night we were fighting. We learned to sleep during the day. This went on for several days. My weapon is a lot cleaner than I am. I threw away my washcloth, a bar of soap and a soap dish to carry more ammo. I loaded my jacket and parka with magazines for my Browning automatic rifle.

After a few days, we were ordered to retreat, a strategic withdraw. Major General Oliver P. Smith remarked: "Retreat, hell! We're not retreating, we're just advancing in a different direction." We charged in a different direction towards Hagaru-ri. During the daylight we moved. At nighttime we would fight. The attacks from the Chinese are still

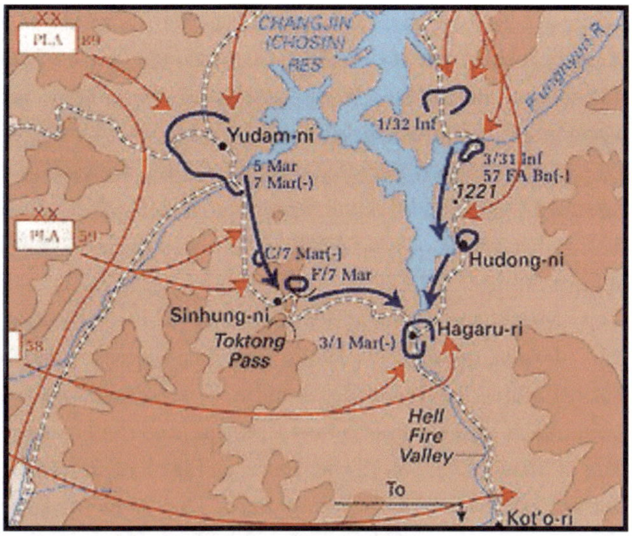

Figure 4.

coming at night while we moved south. Our objective was to get to Fox company who was holding Toktong Pass. The pass was very important for our withdrawal.

No one should have survived. The Chinese had us. Years later, while at a meeting of the Chosin Reservoir survivors in Las Vegas, a former US envoy to China asked me how we survived. He said he had been approached by a Chinese General who wanted to know how the Marines had escaped. He said, "we had them in the palm of our hands." I told him we had exceptional leadership, intelligence, and Marine training.

Fox Company was holding Toktong Pass. I was carrying as much ammo as I could uphill. I had 50 pounds of ammo. It was in the snow and uphill to get to them. The snow stuck to the bottom of our boots. The snow would cling to us, adding to the weight we were carrying. It was very cold, well below zero. My right boot had worn through the sole. My right foot was against the snow. It stills hurts today.

It took us more than two hours to go up the hill to get to Fox Company. The entire platoon going up the hill was maybe 30 guys. It was in the morning when we started. Because of the weather conditions we were in single file about 10 yds apart. A guy from Fox came up to me and told me to give him the ammo and just keep moving. They were out and relieved to see us. At first our officers led us in the wrong direction. Hypothermia effected the feet, the hands, and the brain. We were redirected. I am not sure how many of Fox was left when we got there. We were under fire the whole way up the hill, mostly small arms fire and some mortars. One of the mortars exploded near me and knocked me down. I am shooting my BAR. After we got through the pass the firing stopped for us. But there were others behind us still coming through the pass.

By night fall, about 5 o'clock, it is getting dark and we got to Hagaru-ri. That was our objective. I couldn't walk anymore. I had very bad frostbite and a shrapnel wound. A truck drove by, and I was told to get on it. Had I not, maybe I would not have made it out. They took me to the medic tent, and the medics looked at me. The medics directed me to the plane. I reported to my officer for duty. He told me to get on the plane and get out. He didn't really know how bad my feet were or that I had a shrapnel wound. I also had a concussion from mortars landing near me. One landed so close the explosion threw me a few feet. All of this occurred as I was moving up the hill. It was divine guidance that I made it. And I was flown out from there. Out of my company there was only four of us left. I had bullet holes through my parka. At the mess tent, there was no food left, but I got coffee. I was the second to last to get on the plane. It was a DC-4, I was right by the door. The plane was full and

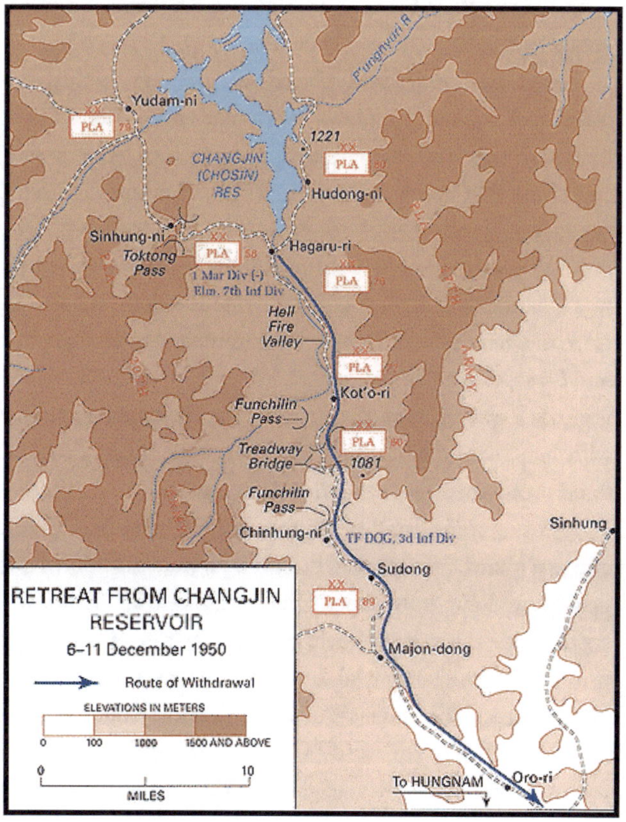

Figure 5.

most were wounded. The plane took us to the US Air Force base at Fukuoka, Japan. Just after we landed, I was put on a stretcher. A nurse gave me a candy bar. I was told to eat it all. I don't remember anything else happening to me there.

I was put on a train and transferred to the US Naval Hospital in Yokosuka, Japan. I was uncomfortable. The seats were hard wood. Many other guys were on the train. It took over 14 hours to get there. First, I was placed on the floor. Then my stretcher was taken into the chapel and placed across the pews. Later I was placed in a ward. They had us strip naked. I remember the nurse carefully picking up our dirty clothes with just her thumb and finger. She added mine to a pile while holding her noise. We had urinated and defecated in our clothes. I got to shower and was given pajamas. We got something to eat and got to sleep. When I got into the bed I started to cry, because my buddy was gone. The guy in the next bunk yelled at me to stop crying. It was my buddy Don Avelar from San Jose. I had been with him since Camp Pendleton. We had lost contact with each other during the battle. Divine guidance again.

A short time later I was sent to Hawaii. I was there a year, mostly guard duty while I was healing. I had a very high classification, so I could be of use around secret stuff. I was at Pearl Harbor Naval Base. I was walking the perimeter of the area where secret stuff was kept. That is what I was told.

On one occasion I placed my rifle up against this building. I was curious about the padlock on a door. I found a piece of wire and I was messing with the lock. A voice said, "what are you doing there trooper?" At the same moment, the lock popped open. I got in trouble. Not too bad. I was only 21 years of age. I went into town on occasion. There was a dance one night. A girl at the dance was from Fortuna, close to where I lived. Her family had moved to Hawaii. She and I spent lots of time together. We went to some Hawaiian festivals together. One time I was at a Hawaiian gathering with her, I was eating some delicious white meat, I was offered more, then I was told it was puppy.

I was transferred to the San Diego Naval Base Stockade as a guard. I was locked in with 64 inmates. Many of the guys were doing hard time. Some awaiting trial. There was a riot inside the prison. I help stopped the riot. Some of the prisoners were put on trial for the riot. I was a witness. Two of the bigger prisoners warned me, "Watch what you say." There was heavy wire and bars over the windows. One of the prisoners had a horned toad. I was wondering how he got it. I took it from him and gave it to my superior. When I went back to this guy, he had a little car, he was playing with it. I let him have it. I may have been there about a month. The threats from those two prisoners got back to my commanding officer. Knowing these two had friends, my commander told me I should transfer.

I didn't know anything about China Lake, but it sounded good. I didn't know it was out in the middle of the desert. I was walking guard duty in the snow. I was there just a few months.

My time was up, and I was discharged. I went home.

I went to work for Pacific Lumber Company. For a time, I was still recuperating from my Korean War wounds; concussion, and frostbite on my feet. I went to the hiring office in Scotia. Doc Nevell was interrogating all of us looking for work. I listened to his questioning. Doc knew me from the day I was born. He asked me, while seeming to look right through me, what do you want to do? I said, "I want the hardest job you have." He said why? I said it pays the best. He asked when can you go to work? I said, yesterday. I started the next day.

Marilyn came into my life when I was 15 in high school. When I got back from my service in Korea, I thought she was probably married. I went to see her. She was as happy to see me as I was to see her. At her parents' Fortuna house, we had our first kiss. She had a fraternity pin on a chain around her neck. I asked her to go with me for a ride. We went out to Centerville. I told Marilyn, "you can't marry this guy." She responded, "Why can't I?" I said, because I love you." I asked her to marry me. She said, "I won't marry a man who hasn't graduated from college." So, I left the Lumber Company.

A friend of mine asked me to go him to Santa Rosa Junior College. I was there for one semester. I had a lot of health issues. The concussion caused terrible headaches. The frostbite on my feet. And I had flashbacks (PTSD). I was very angry, almost anything could get me to fight. I was a hand grenade ready to explode. I did at times. I would rather not talk about it—made me seem like a monster.

A buddy of mine and I hitchhiked to Santa Rosa to go to the junior college there. I had no money. I was starving to death. I met a guy who had been on the football team. He had hurt his shoulder. I was a hunter and had a lot of dear meat. He said to me, "if you do the cooking, I will feed you." I got by. Prior to that all I had was a small hot plate, bread and cheese. I made toasted cheese sandwiches for breakfast, lunch, and dinner. I slept in my car some nights.

At Santa Rosa JC, I had history, English, hygiene, and PE/football classes. I played right offensive and defensive end for the football team. I had great hands. I was only there for the fall semester. I transferred to Humboldt State. I was closer to home. Its 1948. I was thinking about being a coach. I was a good athlete. My best sport was baseball. I had been scouted by the Yankees. I was not consistent enough in my hitting. I was going to teach as well. I wanted to teach art. I like to paint and use charcoal and pottery. My grandkids have all of my old works.

I transferred to Humboldt State and graduated in 1956. My degree was in Education. I became a teacher. Superintendent Oden Hansen of the Arcata Schools said he would hire me as soon as I wanted to go to work. I started teaching in the fall of 1956.

Marilyn was not aware of my anger issues. I never was angry in front of her. She was a calming influence on me. She helped me get through what I was going on inside me.

I was assigned to Sunny Brae. I was teaching 5th grade. About 30 kids in my class. I was there for a year. Then I went to Bloomfield School. I was teaching a special ed class. These kids had problems. They were hoods! Two of the kids were just mean kids. They challenged me to a fight. I had not been long out of the marine Corp. I was a young man. I told them, OK, and started to take off my jacket. I turned my back and they jumped me. I said, wait one second, and faced both of them. They just looked. I said we will put down a pad to protect your head when it hits the floor. Very shortly, I had both of them giving up. Another teacher watched the whole thing. About 10 kids in total. I was there for a year. The next year I was at the Stewart School and had a self-contained 7th grade class. I was there for 8 years.

Our first child Lance was born in 1960. Our second Kirk was born in 1962. We have seven grandkids, one of whom is a teacher.

During the beginning of the 1964-65 school year my headaches became more severe. This was the results of my wounds from the Korean War. Specifically, my concussions cause more problems. I was also having difficulty sleeping. I went to a doctor, then had my eyes checked. Then I was sent to Sacramento to see an eye specialist. He sent me to UC Berkeley. The exam there found a brain tumor on my optic nerve. They did the surgery. During my recovery, the VA got involved. After the surgery I had no vison in my left eye and about 30% vison in my right eye. My surgery was in December of 1964. I was back home in January. Marilyn went back to work in early 1965. I went back to teach in March of 1965.

During the summer of 1965 I transferred to the Eureka schools. Eureka was where we lived. I started to teach history and art at the High School. Art was my major. But my vison started to fail, so they gave me math classes and an aide to help me. I taught there for 20 years with very limited eyesight. The last ten years I could not have done it without my aide, Diane.

One time I was at the blackboard with my back to the class. There was a commotion. My aide was not in the classroom at that moment. I called out, "Harry, knock-it-off." Never turned around. Harry said how do you know it's me. "I have eyes in the back of my head!" At the end of class, Harry came up to me and very politely asked, "Mr. C can I see those eyes in the back of your head?"

One time I was teaching division of fractions. You had to invert the divisor and then multiply. You have to learn the language. Invert means to turn it over. One young man didn't understand. I told him to come forward and had him sit on the floor. I picked him up by the ankles and held him upside down. And I told the class this means invert!

Oden Hansen saw me do this. I got a lot of good help from a lot of good people.

A total of 34 years as a teacher. Unfortunately, Ray passed away on Memorial Day 2020.

California North Coast Vehicle and Human Use Beach Survey of False Klamath Cove

Ruthie A. Maloney (Humboldt State University), John W. Corbett (Humboldt State University)

Keywords: survey, mpa, false klamath rock, tribes, statistical base line data

The California north coast (NC) lacks adequate statistical information on beach use. In an effort to provide current statistical data of human and domestic animal uses of NC beaches, a group of concerned and like-minded people formed the North Coast Native Protectors (NCNP) and the Eagle Eyes of False Klamath Cove (EEOFKC) study was developed. This study is located in the northwest corner of California in the County of Del Norte. The area is rural, and many inhabitants are Native Americans associated with federally recognized Tribes. The town of Klamath is approximately seven miles south of False Klamath Cove and is the headquarters for the Yurok Tribe, 80% of the residents are Native Americans from local Tribes (Tribe). This study consists of people who are aware of the north coast's unique landscape and are willing to contribute their expertise and knowledge to create local statistical information necessary for marine planning and management.

North coast communities include Tribes that practice their cultural traditions to this day as they have since time immemorial. Traditional Ecological Knowledge (TEK) of local Tribes should be acknowledged and included because it directly affects a cultural way of life that has existed since time immemorial. Preserving and maintaining traditional marine resources for future generations is essential to the religious ceremonies, traditional foods, and the health and well-being of California Native American Tribes. Our goal is for Tribal governments, Native Americans, and north coast communities to deliver high quality data for coastal management planning. This study is the first baseline data collection and analysis for the NC at False Klamath Cove (FKC) beach in Del Norte County. This study welcomes all races and Tribes to participate in EEOFKC/MPA Watch and baseline Behavioral Census Survey activities. This is important to end prior exclusions of Native Americans from marine planning science efforts. To implement these goals, we conducted a scientific observational study of False Klamath Cove, which is in the heart of Yurok Ancestral Territory, represents Park visitor populations, is heavily used by regional standards, and has an unusual variety of habitats to observe human behavior. This behavioral study reviewed the kinetic movements by boat, car, drone, bicycle, motorcycle, pedestrian, and domestic animal uses of coastal and marine resources at FKC in Klamath, California.

Observational data collection of human usage in and around FKC began in 2017 by enlisting the help of Tribal members, volunteers, and community groups. The location of the study lies within a designated Marine Protected Area Special Closure (MPASC) for bird rookery protection, False Klamath Rock, and took place from March 1st to August 31st annually (Game). The rookery protects the nesting areas of some 40,000 birds. In February 2018, EEOFKC teamed up with WILDCOAST/Marine Protected Area Watch (MPAW), a statewide collaborative effort to collect human use information to enhance the management and performance of MPA's. Data collected from this collaboration informs management, enforcement, and science of

Figure 1.

California's MPA's, providing insight on how human uses are changing because of MPA implementation. By involving local communities in this important work, North Coast Native Protectors and Marine Protected Area (MPA) Watch programs inspire and empower stewardship and educate California communities about ocean ecosystems (WILDCOAST). Volunteers are trained to observe and collect unbiased data on the uses of coastal and marine resources. This provided the basis of a successful citizen science monitoring program. Community and citizen science (CCS) projects of many kinds have played a prominent role nationally and in MPA monitoring and education, providing significant value to the State of California in many natural resources management policy decisions. CCS is relevant to multiple goals of the Marine Life Protection Act (MLPA) related to ecological performances as well as promoting research and education and even effective enforcement (Meyer).

Stewardship of False Klamath Rock Rookery

Reviewing the historical backgrounds of sites, to avoid massacre areas, is an essential consideration for Native American citizen participation in survey and research within ancestral territory. The FKC survey area covers former village sites by both the Yurok Tribe and Tolowa dee ni Nation. The Yurok village of 'Omen hee pur is located on the North side of FKC near the mouth of Wilson Creek (Tribe). The Tolowa of Crescent City and the Smith River area also claim an ancestral tie to these village sites and call them Daa-ghelh-ts'a'. No massacre occurred at this site and local Tribes would gather once a year for a marine feast and peaceful settlement of disputes. It is traditionally associated as a safe place for Native Americans.

FKC is an ideal site for study due to the wide variety of cultures and habitats. The coordinates are N 41 65249, W 124 10773 located approximately five miles North of the Klamath River. The site has variable substrates. This variety of substrates allows a study of broad-based beach behavior in a variety of different FKC beach habitats. The Northern boundary of FKC is the southern boundary of Wilson Creek, which sometimes moves. Wilson Creek is a freshwater stream with anadromous fish that were traditionally fished by Native Americans with gill and dip nets (Tribe).

The habitats of the site include beach, breakwater, cobblestone and reefs. We visited and appraised each habitat site, and the habitats clearly fit into each category. Beach is defined as loose deposits of sand including some gravel or shells that cover the shoreline (Wiki). Breakwater is defined as a barrier built out into a body of water to protect a coast, harbor, or highway from the force of waves (Merriam Webster). Cobblestone beach is defined as a class of rock defined on the Uddin-Wentworth scale as having a particle size of 64-256 millimeters (2.5-10.1 inches), i.e., larger than a pebble and smaller than a boulder (WIKI). Intertidal are the rocky areas of coastline between the low and high tide marks. This area is from the high-water mark to a depth of 2 feet (Google).

Figure 2.

Methods

The survey was conducted in three phases: (1) testing beach survey and proper survey protocol, (2) a yearlong survey generating sufficient data to determine the frequency for a random survey, (3) a random survey.

Phase One: NCNP conducted a beach survey data search. No previous studies were found. State and National Park Service staff were interviewed on attendance highs and lows as well as key holidays. There was a preliminary collecting of snapshot data of the number of vehicles in the North and South parking lots at 8:15 a.m. and 5:15 p.m. each day for six months. Redwood National Park Service 26 years of vehicle survey reports at Lagoon Creek parking lot were examined. Lagoon Creek is separated visually from FKC by a heavily vegetated area but is located nearby.

A paper survey data quality plan, establishing protocols for entering data, was created before the launch of the study. Phase one served the role of a bench survey and more than three testers were used. An extensive pilot study was conducted. While minimum recommendations are for two weeks (IPA Research Protocols), the FKC survey lasted six months. The survey included all uses of the beach, access points, and experimenting with different survey questions to get input from more than three surveyors. From this six-month period, survey questions were formed and tested by observers for clarity and ease of recording. This process determined what survey questions were necessary to cover all beach uses. Survey volunteer responses provided the experience necessary to author a practical volunteer instruction manual (Maloney). Binoculars as survey devices to enlarge and increase visibility were tested with favorable results.

Through the process of Phase one, we identified survey sites and issues. This was followed up by a transect survey. Consultation was made with MPA Watch, who administered the largest citizen beach survey and enforcement program in the State of California. MPA Watch developed academically reviewed lists of survey questions and, with their permission, many of the questions were adopted by this survey (WILDCOAST).

The survey observation site, located at the center of the beach on the west side turn out from Highway 101, was identified and nicknamed the "perch." This site overlooks the breakwater and provides a commanding view of the entire beach. During this test phase, surveyors were placed in the North Parking Lot (NPL) and the results were compared to perch surveys. It was determined that the perch surveys adequately identified beach attendance and activities surrounding the NPL. A similar determination concerning visibility of the south parking lot from the perch was made with the same numerical results. The Lagoon Creek Parking lot and the Yurok Loop Trail head are not visible from the perch. Surveys were conducted where the Lagoon Creek trail entered the beach and beach use was visible from the perch. The results showed exceptionally low traffic levels on that entrance to the beach areas. It was determined by observation that beach users from the Lagoon parking lot can be detected on the beach. Parking for surveyors was provided at the perch turn out. Communication systems with Park personnel were established (Park). All night surveys were conducted that established the almost complete lack of parking lot and beach use during the night. The survey established high levels of public compliance with posted beach hours of sunrise and sunset. The intensity of high beach attendance days were tested with the survey form, and it was determined surveyors could keep up.

The weather for FKC was determined to be a major factor affecting swimming and water sports. The cove receives 79.09 inches of rain a year over an average of 93 days with .01 or more precipitation. Beach attendance drops dramatically when it starts raining. Additionally, rough seas at FKC and the entire extreme northern California coastline have many days with higher waves than other areas of the state. The average daily high ambient air temperature is 61 degrees (data). The average water temperatures are considered cold for a swimming beach at 11.2 C, and this no doubt discourages water activities (Google).

Phase Two: It was determined that a year of surveying would be required to gather sufficient evidence to properly determine the monthly frequency required for a random survey. There was a development of using an I-pad to instantaneously record results. This effort did not prove successful because reception at the site was unreliable. Reconciliation protocols for paper recording followed throughout the entire multi-phase survey period. Personal communication with Dr. Steven R. Martin, Ph. D. Professor and Department Chair of Environmental Science & Management at Humboldt State University, determined that Phase Two survey parked cars could be used for establishing sampling frequencies on the final Phase random survey. Dr. Mark Rizzardi, a tenured statistics professor at Humboldt State University, reviewed the calculations (Corbett, Attorney at Law).

Phase Three (The official survey): The recruitment of surveyors took place in one of the lowest income areas of California with an extremely high unemployment rate of 30% (EDD). This provided us the opportunity to recruit readily available, educated Native Americans for long 12-hour survey shifts. The effort instilled in the local community participants resulted in a sense of accomplishment and pride in their work ethics and commitment to this study. This study provided skill set training for Native surveyors, many of whom have used their experiences over the last two years to find employment. Our goal has been to broaden the base of Native American participation in supporting healthy oceans. This is the only multi-Tribal citizen beach monitoring study in the State of California and the only one run by Native Americans. Additionally, the EEOFKC/MPA Watch survey is a visual observation from a public area without interaction with the public nor identification of individuals. Therefore, no science or IRB permits were needed.

The frequency established to achieve a 5% accuracy was six days each month plus one extra day for the year for a total of 73 days. Each month six days were randomly selected. In addition, a marked seventh and eighth day were selected. These extra days were used in case a survey was short of hours, and hence discarded, or did not go forward on the planned date because of a surveyor's life circumstances. If a selected date did not happen, the next random seventh and eighth day was selected. The planned extra day,

Figure 3.

number 73, was selected first by a random month and then a selected day. There were three extra days because communications were unclear that a survey had been completed, and the next random days were selected. Rather than throw the surveys out, this resulted in three more days than were required. This was approved by e-mail correspondence with Dr. Martin (Martin).

The total days surveyed was 75, with 36 days in 2019 and 39 days in 2018. The average number of vehicles per day was averaged to develop strata categories. Strata number one was Saturday and Sunday. Saturday had 13 surveys totaling 4,280 vehicles and Sunday 12 survey totaling 4,338 vehicles. Strata two was Monday, with eight surveys totaling 2,131 for a total of 32 survey days equaling 10,749. Tuesday had ten surveys totaling 2,346 vehicles. Wednesday had ten surveys totaling 3,305 vehicles. Thursday had ten surveys totaling 2,906 vehicles and Friday 12 surveys totaling 3,251 vehicles. These numbers provide the basis for calculating total vehicle numbers from which to project a beach census number.

The survey consisted of six-pages that was completed by surveyor's starting at 7:00 a.m. to 7:00 p.m., which included time for breaks and lunch. Some surveyors split the shifts and each person of a team completed two six-hour surveys in shifts from 7:00 a.m. to 12:00 p.m. and 1:00p.m. to 7:00 p.m. At the end of each hour a new survey form was used. The EEOFKC conducted a beach survey data search. No data was found. State and National Park Service staff were interviewed on attendance highs and lows as well as key holidays.

Hourly survey forms of human beach use were collected by recording all offshore and onshore coastal activities within and outside FKRSC. Volunteers were trained to recognize different types of activities, using binoculars to view activities offshore and on shore, and to record what they saw on data sheets. Surveyors began and ended their surveys each hour. All collected data underwent rigorous quality assurance and quality control protocols (WILDCOAST). There were two designated inspectors to ensure surveyor compliance. Periodically, early morning inspections of volunteers occurred at the 7:00 a.m. start time and the 7:00 p.m. quitting time. There was no cell phone service, so surveyors who left for health or family concerns were directed to call one of the inspectors immediately when they entered an offsite cell phone reception area. Some surveys did not meet time standards because they surveyed for the whole 12-hour day and were discarded. Surveyor or personnel changes were made as appropriate.

The California Department of Public Health oversees the issuing of advisory warnings to not consume bivalves, and sometimes crustaceans, if they have dangerous level of Paralytic Shellfish Poisoning (PSP) and Domoic Acid Poisoning (DAP). There is a standard closure for the entire state for each year from May 1 to October 30 every year for many decades, which is for mussels. If conditions warrant the quarantine dates are extended. In recent years, all bivalves, clams, and sometimes crustaceans, such as Dungeness Crab, have been added to the quarantines. The effect of these health quarantines was not studied in this survey. We are not aware of other beach surveys that have quantified the effects. A review of tide pool harvest rates reflected in the survey suggests this will not substantially affect survey conclusions (Health). Most of the PSP baseline data is gathered by the Yurok Tribe for the California Department of Public Health. The Tribe independently notifies Tribal members and locals of closures.

NCNP, EEFKC, and RAM Consulting provided surveyors, a data entry specialist, and analyzed the data. Data entry was subject to a quality control program that required a minimum of three independent reviews of the data. The excel spread sheet was independently developed by an outside contractor. Data entry was independently entered and was then transferred to an independent auditor to review the data. Differences were recorded by survey date, time, and entry. The auditor reconciled the data. Next, the data was reviewed by Executive Director Ruthie A. Maloney, MA. Project advisor, and John Corbett carefully reviewed the data entry process and the results. This procedure met the data quality control plan and a statistical protocol for paper surveys (Service).

The California Center for Collaborative Policy at Humboldt State University conducted the final analysis of data in SPSS Statistic program. SPSS totaled the surveys as to each category, area of the beach, and human behavior for all 2018 and 2019 survey months. Each habitat type, such as intertidal rocky and sandy beach, were totaled. The types of vehicles (Federal Highway Admnistration) of each of the four parking lots were tallied, i.e. Northwest parking lot, Westside (Southbound lane), East (Northbound lane) of Hwy 101, and the South West Parking lot. Total domestic animal use in each section off leash vs on leash were record.

Results

Survey results were then calculated for annual projections of vehicles, human and domestic animal use of FKC beach. This data was compiled in a format acceptable to Humboldt State Professors Dr. Martin and Dr. Mark Rizzardi, who have conducted many census studies (S. a. Martin). The data showed, based on a sample of 86 days at 12 hours a day over a year, that there's an average of 396 vehicles per day, with a standard deviation measure of the day-to-day variation in vehicle counts of 167 cars. In order to achieve a 5% confidence interval (+1-20 vehicles per day) requires a sample of 73 days/year (– 6 days/month) for a sample (Martin).

We determined from the Phase three survey that there were two strata. Strata one consisted of Saturday and Sunday and strata two consisted of Monday through Friday. The strata were selected based upon the survey results and were reduced to average cars by day of the week.

A minimum of six surveys a month were required. For most months, eight days were randomly selected to ensure non-bias. This was in case one of the required six days did not go forward on the planned date or a survey was short an hour and discarded. Road closures and life circumstances of surveyors resulted in the need for the next date of the random draw. Total days surveyed was 75, with 36 days in 2019 and 39 days in 2018. Survey days are in the chart below.

Days of strata were determined by reviewing the chart. Strata one is Saturday, with 13 surveys totaling 4,280 vehicles and Sunday with 12 surveys totaling 4,338 vehicles. Strata two is Monday, with eight surveys totaling 2,131 for a total of 32 survey days equaling 10,749 vehicles for Strata one survey results. Tuesday had 10 surveys totaling 2,346 vehicles, Wednesday had 10 surveys totaling 3,305 vehicles, Thursday had 10 surveys totaling 2906 vehicles, and Friday had 12 surveys totaling 3,251.Out of 892 hourly observations, about half (67%) or n = 579 occurred on a weekday (Monday– Friday) and about 33.1 n = 287 of observations occurred during the weekend (Saturday –Sunday).

From 2018–2019, there were a total of 892 observations documented at FKC. About 460 were from 2018 (July 1, 2018, to December 23, 2018) and 432 were from 2019 (January 1, 2019, to June 27, 2019). The total of these numbers were converted to daily averages and multiplied by the days of the week per year for the total beach and vehicle census data.

Each observation was gathered by individual location and parking sites, making for a complex survey but one where detailed site information can be developed for each sub site. Combined figures are provided for interpretation and analysis. Therefore, the results from the observations will be described by combining all the parking lots. Parking lots were counted in two ways: (1) Redwood National Park and Humboldt Redwood Park ownership sites only, i.e. the Northwest and South parking lots, and (2) the public parking on the sides of the road overlooking the beach. The study area included a Northwest parking lot, South parking lot, West side (South Bound Lane), and East Side (North Bound Lane) of Highway 101.

The survey data shows that a total of 22,486 vehicles were observed and recorded from all four areas. Formula

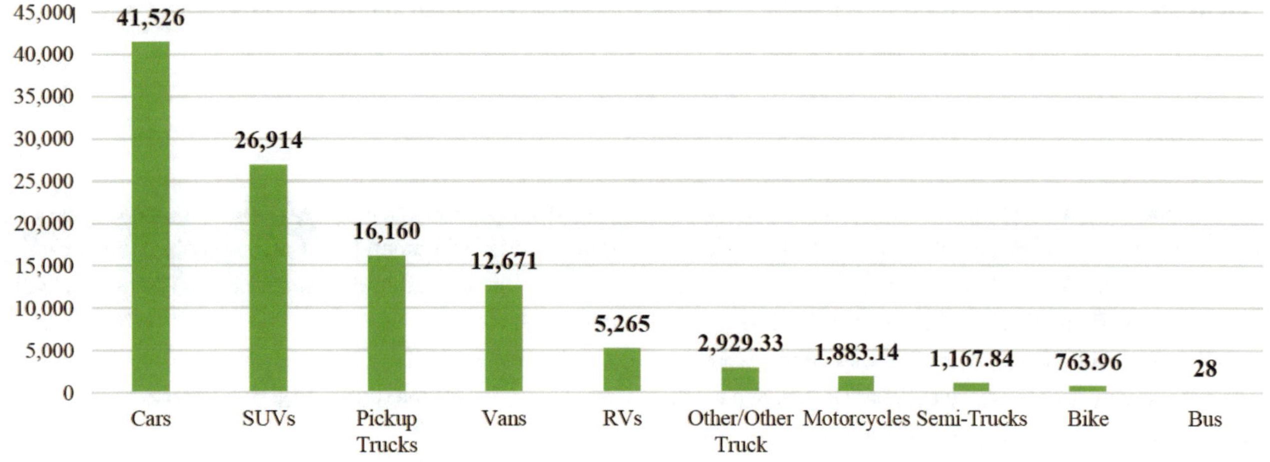

Figure 4. Annual Vehicle Projection at False Klamath Cove by Vehicle Type (n = 109,416.87).

applied 22,486 / 75 survey days = 299.8 cars per day x 365 days a year for a projected total of 109,427 vehicles annually.

To gain further insight into the number of people in the vehicles, this study applied the National Highway Administration (2009) modes of transportation to recreational passenger numbers from a chart provided by Cal Trans.

This study added 2 additional categories not included in the Cal Trans as Highway conversion chart. (1) Bicycles were calculated at one person, and (2) two persons were assigned for each bus, because it is exceedingly difficult to count passenger numbers and no data on how many for bus use was available. This will result in an undercount of the number of bus passengers by vehicles counted in the survey. Of course, bus passengers were counted at 100% if they entered the beach.

Applying the California Average Vehicle Occupancy (Persons) by Mode and TD Vehicle Occupancy (Mean), survey data shows the total estimate of passengers for all vehicles (at all four parking lots) is about 50,275.13 passengers who visited FKC. Formula applied 50,275.13 divided by 75 survey days equals 670.33 people per day visit FKC. Converting the number of visitors 670.33 x 365 per days per year creates an estimated annual projected number of 244,670.45 passengers that will go to FKC parking lots and beach.

There were various types of human beach activities that were recorded. Activities were grouped into the following categories/sub-categories: (1) on the reef, either visiting, collecting, or doing research; (2) shore beach activities such as walking, playing, using the picnic table, hiking, sunbathing, rock climbing, running, making a bonfire, sleeping, camping, making art, and watching wildlife; (3) shore based fishing, bi-

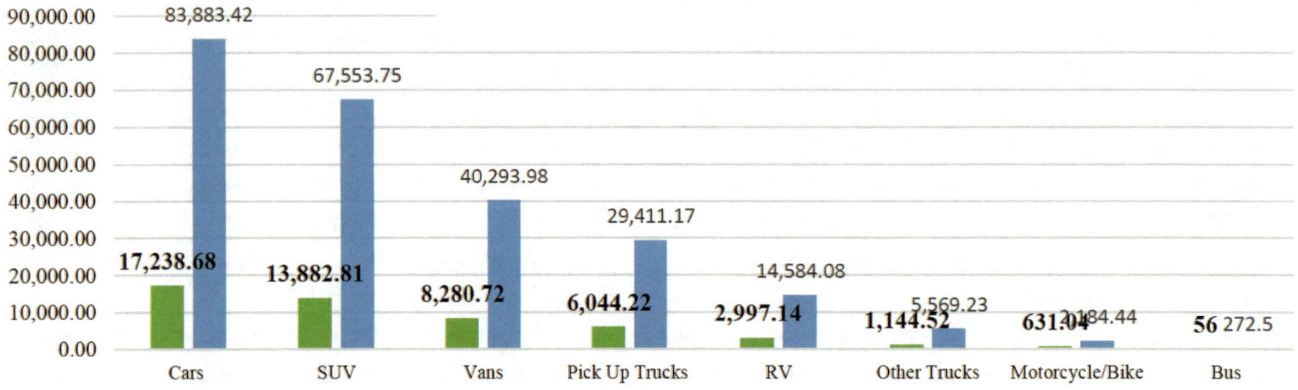

Figure 5. Total estimate of passengers by vehicle type (all parking lots combined and annual projection of passengers. Green bar survey numbers. Blue bar projected annual passengers.

ota collecting activities (interference with wildlife) including carrying a 5-gallon bucket for gathering biota, domestic animal use, hook and line fishing, trap fishing, net fishing, and spear fishing; (4) domestic animal use of beach, (4a) dogs off leash, (4b) dogs on leash; (5) type of clothes visitors wore, including (5a) warm clothes, wet suit, hiking clothes, or a swimsuit; (6) offshore activities including (6a) surfing/boogie board (6b) kite/wind surf (6c) standup paddle, (6d) swimming/body surfing; and (7) type of boats (active/inactive).

Our annual projection is estimated to be 85,363 expected activities to occur at FKC. Figure 6 demonstrates the total number of human-based activities observed, including recorded (green bar) and estimated annual human activities (blue bar).

Domestic Animal Beach Use

EEOFKC/MPA Watch survey showed there were 1,782 domestic animals observed, mostly in the sandy area. Approximately 994 domestic animals were on leash and 778

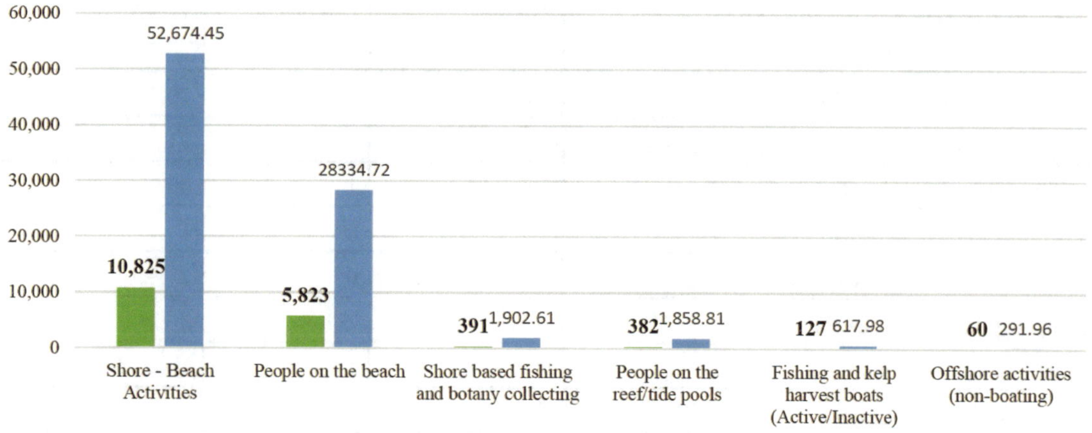

Figure 6. Total human activities at False Klamath Cove by Type.

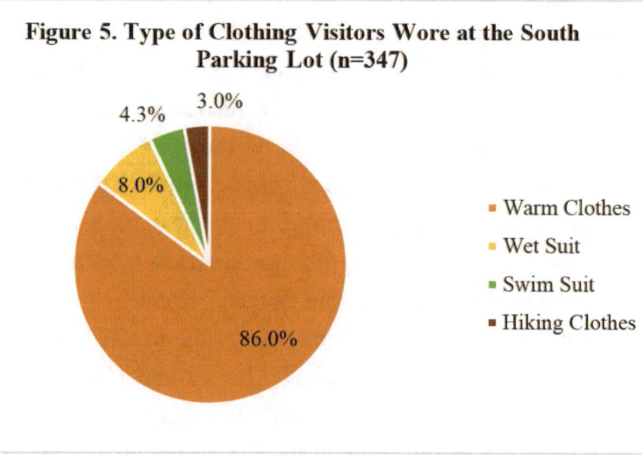

Figure 7. Type of Clothing Visitors Wore at the FKC among 347 visitors.

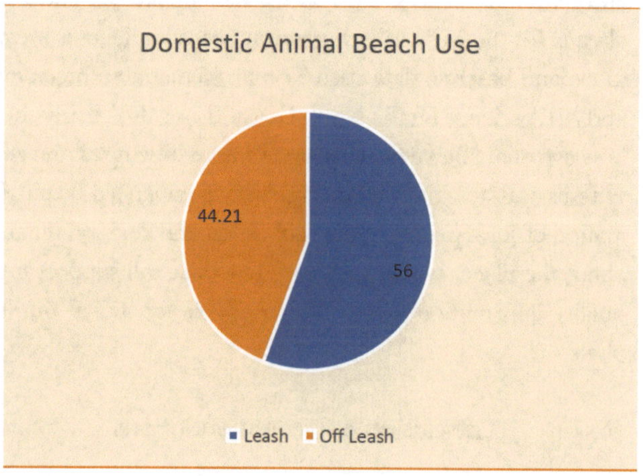

Figure 8. Domestic Animal Beach Use.

were observed off leash. It was noted that people tend to have their dogs off leash in rocky areas compared to on a leash.

Formula applied the total number of domestic animals 1,782 / 75 survey days equal 23.76 domestic animals per day x 365 days a year for an annual projected number of 8,672.40 domestic animals that will visit FKC beach.

Boating

A total of 127 boats were recorded. Figure 9 represents the type of boats observed at the northwest parking lot. There were no fishing boats utilizing a net, trap, or line documented. The low number of fishing boats may reflect the long distances between ports and rough seas.

Outlook

The potential users of EEOFKC and MPA Watch data might include academia, natural resource management agencies, local communities, and Tribes. A key focus for the

Type of Boats Documented at the Northwest Parking Lot	Active	Inactive	Total
Unknown Type of Fishing Boat – Recreational	16	3	19
Unknown Type of Fishing Boat – Commercial	17	4	21
Unknown Type of Fishing Boat – Unknown	79	1	80
Net Fishing Boat – Unknown	1	0	1
Hook and Line Fishing Boat – Commercial	1	0	1
Trap Fishing Boat Commercial	1	2	3
Dive Fishing Boat – Unknown	0	1	1
Kelp Harvest Boat – Unknown	1	0	1

Figure 9. Type of Boats observed and whether they were inactive or active.

program was to inform California's management of MPAs and marine resources. This data is being collected and used already for the MPA Watch program statistics. Data is meant to expand baseline data used by management, enforcement, and MPA science on the North Coast. It can also clearly show how essential Native peoples are to the well-being of the ecosystems and how marine management is enhanced by participation of local people in data collection and decision making about the places where they live. The data will support high quality information that will be available for Tribal marine plans.

Conclusion & Recommendations

The findings from this study can offer important data for future policy related to beach management, public health, land use planning, environmental sciences and economic development. Rural areas are managed based upon available information. The lack of such information has created State natural resource management programs based solely upon Southern California beach use patterns. A positive way to address this problem is developing local data base monitoring. Given the overall lack of funding and research interest, this is a perfect area for citizen science to produce high quality base line data. For a rural beach, False Klamath Cove has significant attendance of approximately 85,681 thousand visitors annually, and we hope data like this will address the lack of natural resources management information.

Marine citizen science programs were developed to be an integral part of marine management and were first developed by using fisherman logs, including Commercial Passenger Fishing Vessel (CPFV) logs, rather than requiring expensive on-board monitoring by scientists. This has grown until now and, in California, major beach monitoring occurs under programs sponsored by the California Resource Agency as part of a statewide program. The University of California Davis is providing support services for such a program. Our project is unique in the State of California because it is being run by Native Americans; 90% of the surveyors are local Indigenous peoples. Unlike many Tribal project's, participants are from many different Tribal memberships: Yurok, Navajo, Tolowa, Karuk and Hoopa. This project is a testament to the capability of Native Americans to participate in the generation of data within their traditional ancestral territory.

The FKC beach site runs itself. Entrance is obtained by turning off the road into two national park service parking lots and then stepping off the lot and walking on the beach. There are no ranger stations or checkpoints. There are bear proof trash receptacles but no retail outlets and no onsite restrooms. Visitors are directed to restrooms that can be found down the next highway.

These restrooms are not visible from the beach. There are two fire pits and one picnic table on the north parking lot. By and large visitors follow the signed rules, except that many dogs run on the beach without a leash. There are occasional Redwood National Park patrols. There are minimal homeless users probably because it is so far from services. There was a homeless person nicknamed the "pipe man" for living in the large culvert beneath the highway. After several months the rains came, and he left. There are rare beach camping fires. Park rules are from sunrise to sunset and closely adhered to. The one exception are families sitting at the lone picnic table who are late barbecuing their dinner. They will wait past closing time to cook and eat their meal. Many of the visitors seem inspired by a reverence for the National Park. Locals have a special relationship with their local beach park as well. From a Native American standpoint, the beach was a traditional annual meeting place for local Tribes to feast together and where problems were worked out peacefully. The primary attraction of the site is the scenic nature of the location.

It is important to remember that the monitoring sites were different between the car surveys and the beach surveys. You had to be on the beach to be counted in the beach survey. For vehicles, the survey area was expanded to cover all four parking sites including bluff sites that overlook the beach without access. Visitors stop here, take a photograph and then drive on. Combining the beach and parking lot sites creates a combined total of 229,775 beach and view sites. This shows that photography and selfies are key recreational uses of the site. While not monitored for, a smaller data set shows 65,628 visitors taking pictures and then driving off and never entering the beach. This suggests the actual numbers are very high and no doubt strongly contributed to the numbers of beach visitors when all parking lots, including non-beach access lots, are considered monitored for. We are unaware of any other beach use survey that includes any quantification of photo/selfies. We are in the process of further refining our survey and expect to find significantly more visitors participating in photos and selfies, as we survey sector by sector until there is an overall reliable quantification system.

Our study hypothesis was that actual beach attendance figures on the North Coast as shown in this study would be much less than the Marine Protection Act Science Advisory Team (SAT) projections. This hypothesis was strongly confirmed by this EEOFKC Beach Census Study. There was a further hypothesis that as information was gathered differences with Southern California beach users would be found. That has proved true. Remarkably for a bird rookery area, there are few birds observed flying around and minimal bird watching. The birds tend to leave the rookery area as soon as they can and disperse along the coast. Mussel and tide pool gathering is extremely limited to the extent it is insignificant. The survey shows public harvesting is primarily in the northern reef. This is fortuitous because the base line studies performed over the years by Humboldt State University, the Partnership for the Interdisciplinary Study of Coastal Oceans (PISCO) monitoring, and MPA follow up to reserve design are all located in the southern reef. This spatial separation minimizes any anthropological disturbances in the science monitoring areas. The results of this study should provide a conceptual framework for the use of local beaches rather than extrapolations from Southern California Beach surveys.

For large segments of the public, their relationship with FKC is visual and involves kinetic movements to take pictures i.e. nature photographs and selfies. There is a national trend of photography but it is particularly evident at FKC. This is due to the growth in technology from facing cameras and the immediacy of social media planforms (Evans). This also seems to be due to an adaptation to environmental forces, such as the cool ambient temperature and cold seas. Two extreme examples of this behavior are visitors taking pictures out of their front windows and then leaving without ever getting out of their car. In another example, a truck driver parked alongside the road and climbed from the cab door straight to his flatbed, took a scenic shot, and then returned to his cab without ever touching the ground. This use of visual beach resources is growing ever larger compared to other more physical beach uses.

Much of the Beach behavior is heavily influenced by environmental factors. These factors are cold ambient air temperatures, cold ocean waters and rough seas. In response to these factors, beach users have moved activities upslope from the ocean, and it is expected this will guide beach behavior on other North Coast beaches. Upslope Beach use at FKC is 86%walking on the beach with warm clothing. This contrasts with a recreational survey at Southern California Beach where walkers were only 5.8% of the activity (Christensen). The 1% contact with the water at FKC is significantly less from the 45% in Southern California waters, where 45% have water contact (Dwight). Sunbathing is an inconsequential use at FKC at 1%, compared to being a major attraction of Southern California beaches at 45%. The colder ambient and ocean temperatures are simply less attractive. Surprisingly, this movement upslope has spilled into the parking lots with photographic activity limiting exposure to the elements.

In conclusion, FKC is a vibrant rural beach located in the heart of the redwoods and Native country. The public has adapted uses up the beach slope and away from the water because of the cool ambient temperature, cold marine waters and rough seas. The two major uses are walking along the beach and photographing the beach experience. It is predicted that as surveys of other rural northern California beaches occur, the public adaptations to these environmental experiences will be similar. This provides a conceptual framework for the management of marine resources far different than those of Southern California beach uses. The project represents a positive step by Native Americans to gather data for the better management of their natural resources. Clear, well-thought-out protocols are necessary for the credibility of such citizen science data bases. Such statistical projections are in no way a substitute for Traditional Ecological Knowledge and other science as the Tribes choose to present. The study, while always strongly supported locally, was initially determined not to be grant fundable at the state level review committee for a variety of reasons, including a lack of confidence that Native Americans had the capacity to meet deliverables. NCNP proceeded with the study anyway without funding during the early years.

Our recommendations for what future studies should look at are as follows: (1) what is the average time on the beach at False Klamath Cove and North Coast beaches? Our hypothesis is that average beach times will be less in Northern California than for Southern California. This would be based on cooler ambient air, ocean conditions and rougher seas in Northern California. (2) The Federal Highway Modes of transportation is a 2016 phone survey of 300,000 people and is considered the gold standard for extrapolating passenger numbers by vehicle type. Information gathered coterminous with the event would be more accurate and more recent information needs to be gathered. The word "recreation" is very broad and information specific to beach uses would be best, and consumer vehicle patterns have changed recently. Our hypothesis is SUVs will be more frequent than in the past and that beach recreational travel will have different numbers of passengers. (3) Technology and culture changes have made the use of beaches for photography purposes a major and growing activity. So far this major use of beaches has not been quantified and is lacking in current surveys. The current survey found, from a narrow spectrum of data, 65,602 photographic uses of the beach, suggesting the numbers of such uses will be quite large and significant. (4) We recommend the provision of sanitary bathrooms at the beach, as well as better directions and marketing about the availability of the Lagoon Creek parking lot for buses going to False Klamath Cove. (5) The Humboldt Redwoods Park and the Redwood National Park might require the redesign and refinishing of the Northwest parking lot, as mitigation of environmental disruption from Cal Trans long term Last Chance Grade construction work . (6)There needs to be work done with Native Tribes to develop informational signage in their native language. Marine Life Protection Act (MPA Watch) is already committed to developing language that, when approved by the Tribes, can be used by the National Park Service.

Acknowledgements

Thanks to my wife, daughter, and son. Thanks to all of the following: Dr. Laurie Richmond Humboldt State University, building citizen, fishing and Native American participation in Natural Resources, Catherine Khulman former Executive Director of the Ocean Protection Council, Dr. Snezanna Levic, University of London for constant encouragement, Troy Fletcher. The Yurok Tribe: Deceased former Executive Director of the Yurok Tribe, scientists Mike Belchik and Scientists Dave Hilliemier. Fishing Groups: North Coast Surf Fish representative Dennis Mayo, Humboldt Fish Marketing Association representative Vivian Helliwell, Jacque Hostler Fish and Game Commissioner and Executive Director of the Trinidad Rancheria. Blue Ribbon Task Force members Meg Caldwell, juris Doctor currently deputy Director Packard Foundation Ocean Conservation and Science, Roberta Cordera Chumash, Indigenous leadership, Mediator and Jimmy Smith former Humboldt County Supervisor (deceased).

Waklow (Thank You) to my dad, mother, sister, daughter, and son. It's honor to advocate for the protection of Indigenous religious, cultural, and environmental rights for future generations. Much appreciation to the Yurok Tribe Awok' Troy Fletcher who would be proud of this work, Yurok Tribal Council and Council support (nothing would be done without these people). To all the surveyors, data entry, data analysis (Humboldt State University, Rural Center for Collaborative Policy), accountants and editors. To all the people who visited False Klamath Cove! We love you! North Coast Native Protectors invite you to join in our work and projects contact Ruthie A. Maloney (707) 502-9155 or northcoastnativeprotectors@gmail.com.

References

Administration, National Highway. Average Vechcile Occupancy by Mode and Purpose. 2009. https://nhts.ornl.

gov/tables09/fatcat/2009/avo_TRPTRANS_WH-YTRP1S.html.

Barclays. "California Code of Regulations ." n.d.

Christensen, J. King, Phillip. "A new Generation's Challenges on the California Coast, Summary Statistics from Beach Intercept Surveys conducted in Southern California in Summer of 2016." Institute of Environment and Sustainability (2016 Summer): 8 Chart Primary reasons people go to beach.

Corbett, John W. Attorney at Law 1973.

Corbett, John W. IRB Not needed for hourly recording of EEOFKC survey Dr. Brater. 2018. email correspondence.

Data, US Climate. July 2021. usclimate data.com/climate/klamath/California/united states.

Dr., Brater. Humboldt State University Institutional Reivew Board. Arcata, CA, 14 July 2017. sb64@humboldt.edu.

Dwigtht, Peter et al. "Beach Attendance and Bathing Rates for Southern California Beaches."

Ocean and Coastal Management (2007): 50 - 19, 87-858.

EDD, California. https://data.edd.ca.gov/en/Labor-Force-and-Unemployment-Rates/Local-Area- Unemployment-Statistics-LAUS-Del-Norte-/jtkp-jzg4/data. n.d. 11-28-2020 November 2020.

Evans, C. "Risky Photography in National Parks: An examination of the role of online identify management in wildlife Risk Perception." Colorado State University , 2018.

Falk, Martin. Impact of weather conditions on tourism demand in the peak summer season over 50 years. City of Encinitas Beach Attendance Report 2016. Encinitas, 2016.

Game, California Fish and. Marine Life Protection Act. Sacramento, July 2010. google. google.com/search=averageocean+temperature+false+Klamath. July 2021.

Health, California Department of Public. n.d.

https://www.cdph.ca.gov/Programs/OPA/Pages/Shellfish-Advisories.aspx.

Maloney, Ruthie MA. "Eagle Eyes of False Klamath Cove MPA Watch Training Manual." July 2019.

Martin, Dr. Interview. John w. c. n.d.

Martin, Steven and Rizzardi, Mark. Personal communications with Humboldt State University faculty 2018-2019.

Meyer, Ryan Dr., Korabik, Angela, Ballard, Heidi Dr. Examining the Role of Community and Citizen Science Projects in Caliofornia Marine Protected Areas. Davis, CA, May 2021.

Park, Redwood National. IRB Permit needed for EEOFKC observational survey John W. Corbett. 2017.

Ramos, Seafha. "Hikelonah Ue Meggetohl In An Ever Changing World." Cooper-Lara, Kishan & Lara, Walt J. Sr. Ka'm-t'em A Journey Toward Healing. Temecula, CA: Great Oak Press, 2019. 85-93.

Ranger, Redwood national Park. Ownership of False Klamath Cove. Orick, CA, 20 July 2021. e- mail.

Schaeffer, R.L., Mendenhall, W., Ott, L. Elementary Survey Sampling, 4th Editiion. PWS-KENT Publishing Company, 1990.

Service, U.S. Fish and Wildlife. How to Develop Survey Protocols, a Handbook . Fort Collins, CO: U.S. Department of Interior, 2013. Version 1.0.

Tribe, Yurok. "Yurok Marine Resources." n.d.

WILDCOAST. "MPA Watch ." n.d.

CSU Research Competition Articles

Exploring the Effects of the Coronavirus (COVID-19) Pandemic on Fishing Communities and Fisheries Social Science Research in California

Samantha Cook (Humboldt State University), Mikayla Kia (Humboldt State University)

Introduction

The coronavirus (COVID-19) pandemic has disrupted daily life for individuals and communities around the world (Kumaran et al. 2021). Reports from the early months of the pandemic indicate that the effects have been and continue to be particularly disruptive for groups that are already susceptible to social and economic changes, including fishing communities (Mulanda Aura et al. 2020). Fishermen experienced fishery shutdowns, interruptions to the global seafood market, and increased health risks amid existing regulatory and environmental challenges (Campbell et al.

2021). As these and other consequences continue to affect fishermen and fishing communities, fisheries social science research can help to understand the socioeconomic dimensions of the pandemic and support community needs by communicating issues and recommendations to advance advocacy efforts (Bennett et al. 2020).

Before the pandemic started, our Project Team was engaged in an ongoing study to gather and communicate information about the health and well-being of fishing communities in California, including impacts from marine protected areas (MPAs). A key goal of this project is to convey fishermen's perspectives about the unique challenges and opportunities their fishing communities are facing to managers and decision makers. Upon beginning data collection, our team recognized that capturing COVID-19 impacts on fishing communities was both timely and necessary. As such, this paper explores the effects that COVID-19 has had on fishing communities and fisheries social science research in California.

Methods

The Project Team is conducting virtual focus groups with commercial fishermen in each of the major California ports. To date, we have hosted 15 focus groups with a total of 73 participants. We anticipate completing the remaining four focus groups in the coming weeks.

For each focus group, three to eight fishermen representing a range of fishing interests gathered via Zoom to provide their perspectives on their fishing community's health and well-being, effects from COVID-19, and impacts from MPAs. The Project Team selected participants based on their awareness of the state of their port and ability to speak beyond their individual perspective, in addition to demographic factors including fishery of participation, level of experience in the fishery, and familiarity with other focus group participants.

The focus group structure led participants through a deliberative process to rate and discuss 20 questions related to environmental, economic, and social well-being, and MPA impacts. Facilitators posed a question and asked participants to rate their community on a five-point Likert scale using Zoom polls. After participants selected their ratings, facilitators displayed the spread of individual data and asked participants to discuss the areas where their scores differed or, in some cases, coincided. This open-ended discussion allowed for the collection of qualitative data to provide context for the quantitative ratings. Following the discussion, participants scored the same question again to see whether the conversation changed any individual ratings and shifted the group toward a more consensus-based or collaborative rating. These second scores are considered the final rating for the port's fishing community.

Results

We received a total of 51 quantitative responses related to COVID-19 impacts across 11 ports (Table 1). An overwhelming majority of respondents reported that COVID-19 was highly disruptive to their port's fishing operations (Figure 1). One Orange County fisherman explained: "The COVID thing basically killed us, [. . .] that was a kiss of death. As soon as I saw that on the TV, I texted [name redacted] and I said 'we're in trouble,' and the buyers basically quit coming down [to port] a week later. That was it. They were done." Fishermen out of Eureka experienced similar market effects: "We were looking at a situation where there [were] going to be no buyers for our seafood products. We were getting ready to go salmon fishing while the crab price was tanking and our buyers wouldn't even commit to buying one load."

Nearly a quarter of respondents indicated that COVID-19 did not have much of an effect for fishermen in their port (Figure 1). Some experienced new market opportunities, as did this Shelter Cove fisherman: "When rock cod season opened up, people weren't going into town and going to grocery stores; they wanted to stay home. So all I had to do was text a few people and we would have our orders before we even went fishing so we knew how much to catch." While similar responses to the pandemic helped to offset losses elsewhere, nearly all participants felt the economic impacts of restaurant closures following statewide social distancing measures.

Discussion/Conclusion

Preliminary findings reveal that ports across California have unevenly experienced the pandemic and fishermen have relied on an array of creative adaptation strategies.

Table 1. Number of participants per focus group who responded to the COVID-19 question.

Port/Port Group	*Number of Respondents*
Trinidad	3
Eureka	7
Shelter Cove	4
Fort Bragg/Albion	5
San Francisco Area Ports	4
Princeton - Half Moon Bay	7
Santa Cruz	5
Moss Landing/Monterey Bay	4
Morro Bay - Port San Luis	4
Ventura/Channel Islands Area Ports	3
Orange County Area Ports	5
Total	51

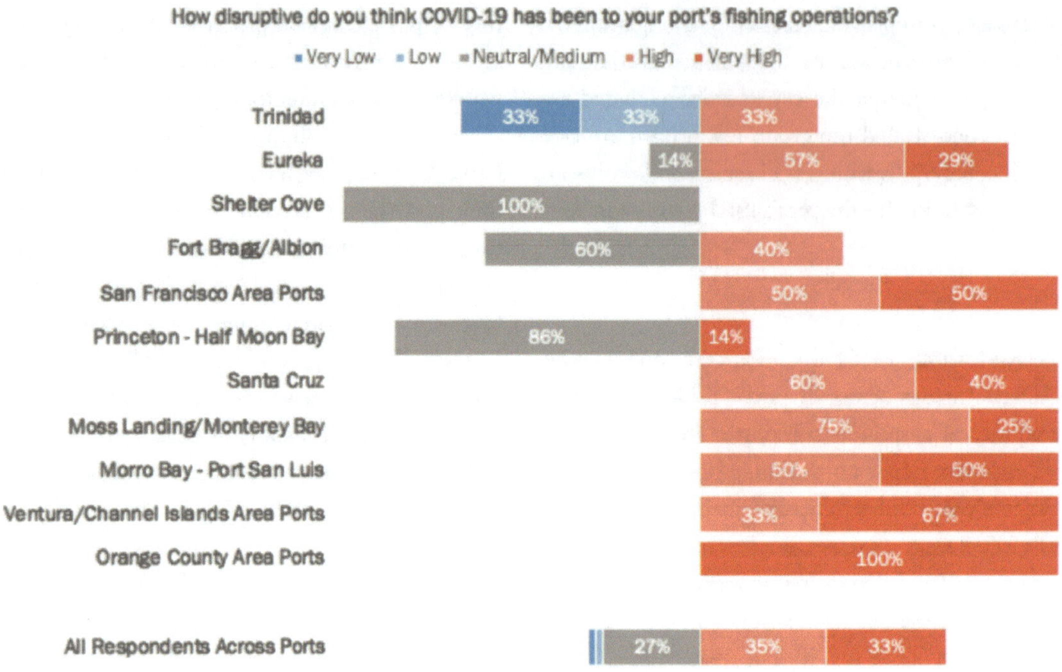

Figure 1. Distribution of responses to the COVID-19 question across focus groups.

For ports that depend heavily on local food retail markets, fishermen turned to direct to consumer sales, often selling their product on the docks. Others, like in Santa Barbara, leveraged their position as seafood suppliers to give back to their communities and donate what they could not deliver to market to local food banks. Similarly, while uncertain market opportunities initially deterred fishermen from fishing, many returned to the ocean either out of necessity or because their job gave them some sense of normalcy. Evidently, continuing fisheries social science research is crucial during this disruptive time, and our team has learned several lessons about ways to continue engagement with fishermen through virtual approaches, including sending meeting materials in advance, hosting Zoom training and orientation, and providing compensation for their time.

References

Bennett, N. J., Finkbeiner, E. M., Ban, N. C., Belhabib, D., Jupiter, S. D., Kittinger, J. N., … & Christie, P. (2020). The COVID-19 pandemic, small-scale fisheries and coastal fishing communities. *Coastal Management*, *48*(4), 336–347.

Campbell, S. J., Jakub, R., Valdivia, A., Setiawan, H., Setiawan, A., Cox, C., … & Box, S. (2021). Immediate impact of COVID-19 across tropical small-scale fishing communities. *Ocean and Coastal Management*, *200*, 1-10.

Kumaran, M., Geetha, R., Antony, J., Kumaraguru Vasagam, K. P., Anand, P. R., Ravisankar, T., … & Vijayan, K. (2021). Prospective impact of corona virus disease (COVID-19) related lockdown on shrimp aquaculture sector in India – a sectoral assessment. *Aquaculture*, *531*, 1-7.

Mulanda Aura, C., Nyamweya, C. S., Odoli, C. O., Owiti, H., Njiru, J. M., Otuo, P. W., … & Malala, J. (2020). Consequences of calamities and their management: The case of COVID-19 pandemic and flooding on inland capture fisheries in Kenya. *Journal of Great Lakes Research*, *46*, 1767–1775.

Best-Sellers: How Our Concerns Translate into Modern Fiction

Joan Esquibel (Humboldt State University)

The mythic mind is well developed in our society; it is the source of expression and inspiration. From this we create stories: films, plays, novels. Together, these form an "imaginative world…that is not just a collection of interesting stories," but "is, more importantly, a way in which participants engage their world." Identifying these stories provides the scholar with a more nuanced understanding of the community's cosmology, its living myths. Myth's relevance is maintained through its flexibility in supporting a variety of interpretations. Over time, this allows the reader "to make sense of the world… its meaning or possibilities."[1]

And those possibilities are flexible too. Mythic narratives are integrated into the community's life, and so they change. Each myth adds to a "continuously produced and reproduced"[2] society, shifting along with the community. The range of any contemporary moment's characters – their habits and conflicts – gives readers an arsenal of behavior patterns to emulate. In times when readers are uncertain of a situation or how to behave, this arsenal provides a persona or mask to act through. And which masks are most commonly adopted can tell us something about a community, and its concerns. Each functional myth provides hints of "literary tastes or social trends for a given period,"[3] making each one a time capsule to the concerns, aspirations, and relationships that readers have with their communities.

Religious myth traditionally filled this role. Today, secular stories allow for social cohesion and common language for people to share within diverse spaces. And these secular stories fill a need. In their 2019 Religious Landscape Study, the Pew Research Center found that a full 26% of Americans identify as religiously "unaffiliated."[4] As religious myths have in modernity started to lose their authority, the modern myths of novels and other popular narrative forms have become our "primary strata of cultural reflection,"[5] and looking at them tells of our future habits.

Through this project I will be looking at the six books that placed the highest on the New York Times Best-Seller list between 2015-2020, to understand the teachings contemporary Americans seek and the morals they perpetuate. Best-seller lists have been criticized for undermining "the book review by not being in the best intellectual interest of the reader,"[6] but that highly academic position itself undervalues the reality of these books' impact on and importance to the general public.

Besides being authoritative, the *NYT* list has the advantage of being representative, in that it is focused on the United States. Going through the data, I noted the book that held the number-one slot for every week from 2015 through 2020. The book that had the greatest number of (not necessarily consecutive) number-one positions in each year, was that year's representative. The representative books are these: 2015 *The Martian* by Andy Weir, 2016 *Me Before You* by Jojo Moyes, 2017 *The Women in Cabin 10* by Ruth Ware, 2018 *Crazy Rich Asians* by Kevin Kwan, 2019 *The Tattooist of Auschwitz* by Heather Morris, and 2020 *Little Fires Everywhere* by Celeste Ng. After reading, the three topics that stood out were wealth, perseverance, and death. This project explores how these themes are portrayed and integrated within the

1 Randall Reed, "Secularism and Myth," *Method & Theory in the Study of Religion* 26, no. 2 (2014): 203.

2 Peter Berger, *The Sacred Canopy: Elements of a Sociological Theory of Religion* (Garden City: Doubleday & Company, Inc., 1967), 6.

3 Laura Miller, "The Best-Seller List as Marketing Tool and Historical Fiction," *Book History*, vol. 3 (2000): 286.

4 "Religious Landscape Study," Pew Research, accessed March 25, 2021, https://www.pewforum.org/religious-landscape-study/.

5 William Doty, "Mythophiles' Dyscrasia: A Comprehensive Definition of Myth," *The Journal of the Academy of Religion 48*, no. 4 (December 1980): 532.

6 Laura Miller, "The Best-Seller List as Marketing Tool and Historical Fiction," 288.

stories, which together act as a representative portion of our mythic canon for this time period.

Wealth is an unsurprisingly contentious theme. In all six books, the narrative perspective is of someone who lacks a resource. Two of the books are set within extreme circumstances – Mars and Auschwitz – while the other four are set in a capitalistic frame where the main character sees but does not have incomprehensible wealth. This suggests that their audiences, who have the resources to purchase these books, crave a narrative that looks at wealth within their society. Within a context of wealth, these stories present themes of gratitude for and acceptance of one's financial standings.

The books simplify society's economic stratification, championing the financially struggling and the poverty-stricken as if all are equally adversaries to the rich. The characters try to comprehend the wealth of their counterparts and after their interactions, the protagonists establish a point of view which dehumanizes the rich in similar ways to "a zoologist forensically examining some strange new creature and its habitat."[7] They develop an emotional detachment from this lifestyle, as if it were only a view "into a life no regular person would ever be able to afford,"[8] and not a possible reality.

The books make apparent a (perhaps surprising) dichotomy between the suffocating and stagnant nature of money and the open and fluid possibilities of poverty. In its apparent perfection, the rich household seems to have sacrificed its warmness, and is described as "a doll's house, where everything is...slightly off-kilter."[9] The young who grow up in these environments are envious of the freedom of poverty, and note that "the kindest people… the most caring, the most sincere"[10] are people in poverty. It is only the poor who can authentically "love their family… [and] feel a deeper sense of pride in who they are as individuals."[11] The books clearly align the readers with those who are not wealthy, and the qualities attributed to the characters serve to uplift readers, who now know that we are perceived in such a positive manner. The wealthy experience murder, infidelity, and harassment, which all point towards a sacrifice of humanity for money. The poor, on the other hand, experience freedom, love, and authenticity, which we are told are the true components of our humanity.

Still, being poor is hard. The amount of resilience in these stories is quite normal – there is a lot of it. The constant, universal, and uniting factor of life is that there are uncontrollable variables that we must learn to interact with. Methods of coping are found within myth, and each story of success provides a further development of these methods. What we learn from these six books are the methods of coping that resonated with readers during this period. The methods that are most prevalent in the texts are these: perseverance through self-confidence and acceptance, overcoming through immersion in work, and that timeless coping mechanism, distraction.

In more than half of the texts, the protagonist's loss of control inspires them to invest in the components of their life they still have control over. Their stories do what religion traditionally has, insisting "that we… aren't helpless, and haven't given up."[12] The past year, 2020, though, has made it especially clear that resilience sometimes looks more like coping, and coping sometimes looks like escapism. An escape does not need to become mindless, but it does alleviate the pressure to make personal decisions and it allows for a deeper examination of issues. So, when readers enter the world of *Little Fires Everywhere*, they escape to the controlled environment of a suburb, but from there, they encounter a complex storyline revolving around wealth, motherhood, death, and race. The distraction of novels does not disregard current issues, and current issues require a certain type of story. None of the books treated in this research project, read alone or separately, could aid us in understanding our pandemic realities, and from that we can see how these books reflect their time. As we move forward, the novels that will be popularized will relate to a feeling of living during and after a pandemic.

Though its presence may increase in post-pandemic narratives, death is a consistent topic. The Best-Sellers do not coach us through our own death. What they do, though, is teach us how to be bystanders to the dying, and one of the more consistent warnings that they present is humbling: demonstrations of toxic and non-constructive positivity do not help the dying. What the stories teach us is that to be an effective onlooker is to listen. For most of the characters facing death, what they most desperately want is to be "reconnected

7 Jojo Moyes, *Me Before You* (New York: Penguin Books, 2012), 58.

8 Ruth Ware, *The Women in Cabin 10* (New York: Scout Press, 2016), 47.

9 Ibid. 56.

10 Celeste Ng, *Little Fires Everywhere* (New York: Penguin Books, 2017), 322.

11 Kevin Kwan, *Crazy Rich Asians* (New York: Anchor Books, 2013), 481.

12 Jared Diamond, *The World Until Yesterday* (New York: Penguin Books, 2012), 347.

with mankind before [they] die,"[13] to have one chance to see a loved one "in this room, just for a moment."[14] They do not seek their loved ones for advice, but just to have someone. The books bring readers into an active, if imaginary, evaluation of the role they play in the deaths of their own loved ones.

The topics discussed – wealth, resilience, and death – are themes we always have engaged with on a mythic level. As an entertaining and accessible mode, best-sellers offer the platform to shape society using the groundworks of religion's mythic systems that our minds have already been trained to internalize, and in its internalization is transformation. This ability, lead towards unity in our society through a shared attitude inspired of these popular novels – the very same novels often dismissed by scholars. As a mechanism that allows for "coherent cultures,"[15] a shared myth is recognizable and sacred to a people over time, which can stem from best-sellers. Myth teaches us to live well by recognizing our aspirations, fears, and goals. And so, as members of a community it is important to engage with these mythic narratives, and as scholars it is important to give credit to the influence of the best-sellers.

Bibliography

Berger, Peter. *The Sacred Canopy: Elements of a Sociological Theory of Religion*. Garden City: Doubleday & Company, Inc., 1967.

Diamond, Jared. *World Until Yesterday: What Can We Learn from Traditional Societies*. New York: Penguin Group, 2012.

Doty, William Doty. "Mythophiles' Dyscrasia: A Comprehensive Definition of Myth." *The Journal of the Academy of Religion* 48, no. 4 (December 1980): 531-561.

Kwan, Kevin. *Crazy Rich Asians*. New York: Penguin Random House, 2016.

Miller, Laura Miller. "The Best-Seller List as Marketing Tool and Historical Fiction." *Book History*, vol. 3 (2000): 286-304.

Morris, Heather. *The Tattooist of Auschwitz*. New York: Harper Collins, 2018.

Moyes, Jojo. *Me Before You*. New York: Penguin Press, 2012.

Ng, Celeste. *Little Fires Everywhere*. New York: Penguin Group, 2017.

Pew Research Center. "Religious Landscape Study." Accessed March 25, 2021. https://www.pewforum.org/religious-landscape-study/.

Reed, Randall. "Secularism and Myth." *Method & Theory in the Study of Religion* 26, no. 2 (2014): 202–10.

Rue, Loyal. "Redefining Myth and Religion: Introduction to a Conversation." *American Theological Library Association* 29, no. 3 (September 1994): 315-319.

Ware, Ruth. *The Women in Cabin 10*. New York: Scout Press, 2016.

Weir, Andy. *The Martian*. New York: Ballantine, 2011.

Research Graphs

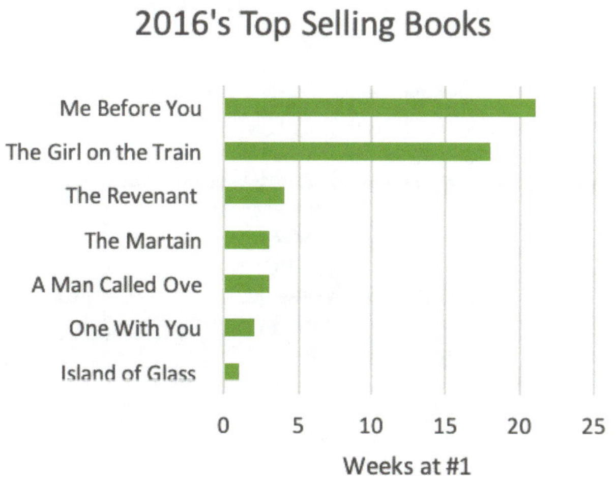

13 Andy Weir, *The Martian* (New York: Ballantine Books, 2014) 100.

14 Ruth Ware, *The Women in Cabin 10*, 268.

15 Loyal Rue, "Redefining Myth and Religion: Introduction to a Conversation," *American Theological Library Association* 29, no. 3 (September 1994): 317.

Research Graphs (Continued)

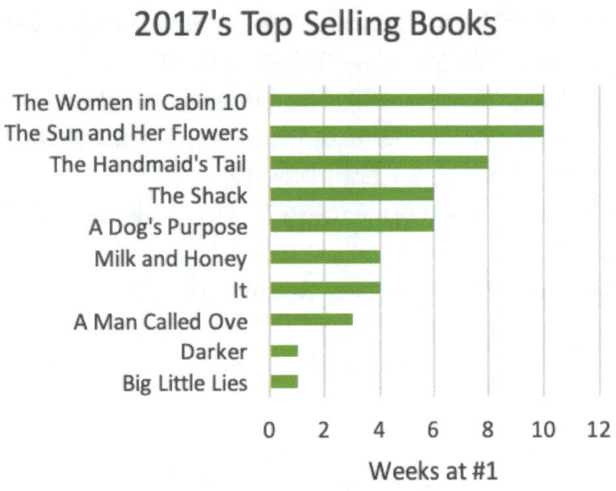

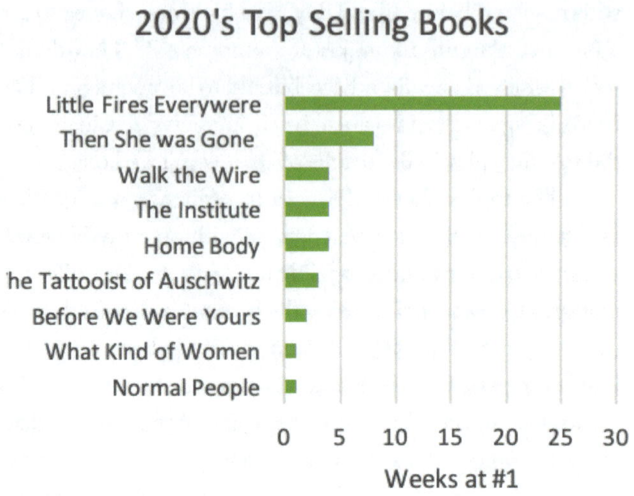

Please view the author's video presentation on this article at https://digitalcommons.humboldt.edu/csucompetitionvideos/2/

Wingate Muscular Power Test Reference Values for Active Healthy Adults Ages 19-35: Normative Data and Differences Between Sex

Levi Basist (Humboldt State University)

Abstract

The Wingate Muscular Power Test (WMPT) has existed for several decades and to this day is considered the gold standard for evaluating an individual's anaerobic capacity. The WMPT can be applied to multiple exercise fields as a means to measure fitness and changes in performance. It is also safe, only requires relatively common testing equipment, and it can be performed by a technician with minimal training required. However, the utility of this test is predicated on having accurate and relevant normative data with which to compare individual results. At present, the existing literature on this subject is either several decades old or inclusive of only specific subject groups. In this study, we seek to record the WMPT results for a large and diverse population of college-age subjects (19-35) in order to generate a new set of normative data. These data sets will include peak power (PP), mean power (MP), fatigue index (FI), and relative power (RP) based on body mass (BM). Our data can then be compared to the previous reference values to evaluate the consistency of the results in addition to comparing performance differences between sexes.

Introduction

Having a reliable method to test an individual's maximum anaerobic power is essential for assessing fitness level, changes in muscular strength, and an individual's capacity to perform in sports or exercises that require powerful bursts of activity (Brown and Weir, 2001). The Wingate Muscular Power Test (WMPT) was developed in the 1970s by the Wingate Institute in Israel (Ayalon et al., 1974). In addition to peak power (PP), the WMPT can also be utilized to measure an individual's relative power (RP) based on their body mass (BM), mean power (MP), and their rate of muscular fatigue (known as fatigue index, FI). Since its inception, it has proven to be a reliable as well as valid assessment and thus is regarded as the gold standard for measuring anaerobic power (Dotan & Inbar,1977; Harvey et al., 2017). The procedure is safe, non-invasive, requires minimal training to administer, and the equipment utilized (cycle and arm ergometer) are already commonly found in most exercise science facilities (Brown and Weir, 2001).

Due to the relative nature of anaerobic power results, it's critically important to have accurate normative data in order to compare an individual's results to that of an analogous population (Hoffman, 2006). A large portion of this exercise research is done at universities with the subjects of those studies often being students in attendance (NSF, 2017). Therefore, it stands to reason that having normative data from a robust sample of male and female college-age participants (both athletes and non-athletes) would provide a substantial aid to future research and analysis of collegiate athletic performance. Several studies have sought to fill this gap in the literature (see Table 1), but oftentimes their sample sizes and/or population demographics prove to be highly specific and therefore limited in applicability. Past publications on WMPT norms have primarily focused on collegiate athletes (Zopan et al., 2009; Baker et al., 2011). Research by Coppin et al. (2012) found reference values from 77 male college athletes. While useful for comparing results with other male collegiate athletes, the normative data lacks applicability to non-athletes and female participants. Other studies have focused on collecting data specifically from female college athletes (Baker et al., 2011). Lack of ethnic diversity has also been observed as a limitation of some WMPT research (Ramírez-Vélez et al., 2016).

Another potential issue with the current research data has to do with the procedures by which the data was collected. Certain testing conventions have been widely adopted, but no universal procedure exists yet (Brown & Weir, 2001). Some of the testing procedures in question relate to which calculation method to use when applying weighted resistance to record PP (Hermina, 1999; Vargas et al., 2015). Another consideration in the WMPT procedure that can vary between testing protocols is the method by which the weighted resistance is applied (Robergs et al., 2015). Other considerations such as cycle ergometer revolutions per minute (rpm) and the flywheel kinetic energy must be taken into account due to their effect on PP output (Bassett, 1989; Hermina, 1999).

Based on a review of the existing literature, the two primary areas of concern for WMPT normative data relate to the sample populations and the application of standardized testing procedures. With regards to the sample population, several studies have sought to fill this gap in the literature, but oftentimes their sample sizes and/or population demographics prove to be highly specific and therefore limited in applicability. Additionally, normative data is only useful when compared to results produced from the same testing procedures. In order to generate normative data that could be utilized widely, but also maintain high standards of accuracy and reliability, this study analyzed and referenced other research procedures and followed the most efficacious testing guidelines currently available.

The purpose of this study was to collect, analyze, and share WMPT normative data that accurately reflects a broader collegiate population than has previously been published while additionally utilizing the most reliable and valid data collection methods. The subject population of this study to collect WMPT normative data includes both sexes, varying athletic abilities, diversity in age, and is non-ethnically homogenous.

Table 1. Comparison of existing WMPT normative data.

Author	*Resistance Level*	*Subjects (Male/Female)*	*Type of Participant*	*Age range of Participant*
Maud, 1989	7.5	62 / 68	Physically Active	College Age (18-32)
Zupan, 2009	7.5	1,374 / 211	College Athletes	College Age (18-25)
Coppin, 2012	8.5	77 Males	Power Athletes	College Age
Ramírez-Vélez, 2016	7.5	1,177 / 667	Healthy Adults	Adults (20-80)

Methods

Experimental Approach to the Program

The WMPT is a commonly utilized method for collecting and evaluating anaerobic power performance. However, in order for these evaluations to be interpreted and valid, the normative data with which it is compared must reflect a robust and diverse sample population and utilize proper testing methods. With these testing considerations in mind, the normative data collected for this study was obtained from a sample of 309 subjects (121 females; 188 males) while performing a standardized WMPT utilizing a resistance based on a percent of the subject's BM. This study was conducted in the Human Performance Lab at Humboldt State University. The standards of 7.5% of BM for females and 8.5% for males was determined based on the established experimental procedures (Table 1). Participants were recruited and tested from December 2015 until May 2019. Measurements of PP, MP, RP, and FI were all collected. Additionally, three RP categories were utilized. The first RP measurement was simply based on the power to mass ratio (PP / BM). The second RP measurement was based on the classic formula (PP / BM $^{2/3}$) which is less biased against heavier athletes (Haff and Triplett, 2016). The third and final RP measurement was calculated utilizing the subject's mean power (MP / BM $^{2/3}$).

Subjects

A total of 309 active healthy volunteers (188 male; 121 female) between 19 and 35 years of age were recruited for participation in this research. Many subjects participated in club or recreational sports, but not college varsity sports such as football, soccer, track and field, etc. All subjects regularly participated in moderate or strenuous exercise for a minimum of 3 days per week for a period of at least 4 weeks prior to participation. Participants were screened for cardiovascular and musculoskeletal disease using a medical history questionnaire, an activity questionnaire, and the Physical Activity Readiness Questionnaire (PAR-Q). Subjects were asked about and subsequently were excluded from the study if they were found to have two or more cardiovascular risk factors as outlined by the American College of Sports Medicine (2013). Subjects were also asked about their use of ergogenic supplements (e.g. pharmacologic aids and/or dietary supplements) that could affect their exercise performance and were excluded from the study if they regularly used them. This study was approved by the Humboldt State University Institutional Review Board, and subjects were informed of the risks and benefits of the investigation prior to signing an informed consent form to participate in the study.

Procedures

Each subject was instructed to complete a general warm-up session following self-paced running for five minutes on the treadmill and dynamic stretching focused on large muscle groups in the lower limbs. Additionally, subjects performed a familiarized submaximal cycling session with 1kg resistance at 50 rpm, including a pair of five second maximal sprints on the Monark cycling ergometer (Model 894Ea, Monark, Sweden). After completing the warm-up session, subjects rested for five minutes before the actual data collection commenced. The resistance was set at 7.5% of body mass (kg) for female and 8.5% of body mass for male subjects. Before performing the WMPT, bike fit (i.e., handlebar, saddle height) was checked and the appropriate resistance was set up on the equipped basket of the cycle ergometer based on subjects' sex and body mass in kilograms. All subjects were instructed to pedal as fast as they could for 30 seconds and remain seated on the saddle throughout the test. At the beginning of the WMPT, the weighted basket automatically dropped when subjects' cadence reached 110rpm. Then data collection began and ran for 30 seconds. Verbal encouragement was provided by the research team throughout the duration of the test. Data were recorded using the Monark Wingate Software (Monark Anaerobic Test Software Version 3.2.1.0) following four main variables: PP, MP, RP, and FI. After data collection concluded, subjects were instructed to remain seated and pedal at a lower resistance for five minutes as a cooldown phase.

Statistical Analyses

The normal distribution of the data will be verified using a Kolmogorov-Smirnov test. Anthropometric data, absolute peak and mean power (Watts), relative and peak and mean total work (watts/BM, Watt/BM2/3), and fatigue index (% decrement of power) will be reported as mean ± standard deviation (SD). All data will be analyzed separately to provide percentile values for males and females. The descriptive statistics will be calculated in mean, standard deviation, and their ranges. A t-test for independent means will be used to verify the differences between males and females. Significance will be set to $p < 0.01$ for all tests.

Acknowledgements

I would like to thank my committee chair, Dr. Young

Sub Kwon, for his incredible assistance in this whole process and for his contributions to data collection. The graduate coordinator, Dr. Taylor Bloedon, is also deserving of great appreciation for her support and advice throughout the entire process. Finally, I would like to express appreciation for the time and effort of my committee members. Thank you all!

References

ACSM. ACSM's guidelines for exercise testing and prescription. Philadelphia, Lippincott Williams & Wilkins, 2013

Ayalon A, Inbar O, Bar-Or O. Relationships among measurements of explosive strength and anaerobic power. In: Nelson RC, Morehouse CA, eds. International Series on Sport Sciences. I: Biomechanics IV. Baltimore, MD; University Park Press; 1974:527-532.

Baker UC, Heath EM, Smith DR, Oden GL. Development of Wingate Anaerobic Test norms for highly-trained females. J Exerc Physiol Online 14: 68–79, 2011.

Bassett, D. (1989). Correcting the Wingate test for changes in kinetic energy of the ergometer flywheel. International Journal of Sport Medicine, 10, 6, 446-449.

Brown, L.E. and Weir, J.P. (2001) ASEP Procedures recommendation I: Accurate assessment of muscular strength and power. Journal of Exercise Physiology Online, 4, 1- 21.

Coppin, E., Heath, E. M., Bressel, E., & Wagner, D. R. (2012). Wingate anaerobic test reference values for male power athletes. International journal of sports physiology and performance, 7(3), 232-236.

Dotan, R., & Inbar, O. (1977, January). 30-sec all-out ergometric test-its reliability and validity for anaerobic capacity. In Israel Journal of Medical Sciences (Vol. 13, No. 3, pp. 326-327). 2 Etzel St. French Hill, Jerusalem 97853, Israel: Israel Journal Med Sciences.

Haff, G., & Triplett, N. T. (2016). Essentials of strength training and conditioning. Champaign, IL: Human Kinetics.

Harvey, Leonie et al. "The Effect of Previous Wingate Performance Using one Body Region on Subsequent Wingate Performance Using a Different Body Region." Journal of human kinetics vol. 56 119-126. 12 Mar. 2017, doi:10.1515/hukin-2017-0029

Hermina, W. (1999). The Effects of different resistances on peak power during the Wingate anaerobic test.

Hoffman, Jay. *Norms for Fitness, Performance, and Health*. Human Kinetics, 2006.

Maud, P. J., & Shultz, B. B. (1989). Norms for the Wingate anaerobic test with comparison to another similar test. Research Quarterly for Exercise and Sport, 60(2), 144-151.

National Science Foundation - Where Discoveries Begin. (2017, November 01). Retrieved October 21, 2020, from https://ncsesdata.nsf.gov/herd/2016/

Ramírez-Vélez, Robinson1; López-Albán, Carlos A.2; La Rotta-Villamizar, Diego R.1; Romero-García, Jesús A.1; Alonso-Martinez, Alicia M.3; Izquierdo, Mikel 1,3 Wingate Anaerobic Test Percentile Norms in Colombian Healthy Adults, Journal of Strength and Conditioning Research: January 2016 - Volume 30 - Issue 1 - p 217-225

Robergs, R. A., Kennedy, D., Gibson, A. L., Zuhl, M., Hsu, H. S., Beam, J., ... & Estrada, E. (2015). Evidence for the invalidity of the Wingate test for the assessment of peak power, power decrement and muscular fatigue. Central European Journal of Sport Sciences and Medicine, 10(2), 63-78.

Vargas, N. T., Robergs, R. A., & Klopp, D. M. (2015). Optimal loads for a 30-s maximal power cycle ergometer test using a stationary start. European journal of applied physiology, 115(5), 1087-1094.

Brown, L. E., & Weir, J. P. (2001). ASEP procedures recommendation I: accurate assessment of muscular strength and power. Journal of Exercise Physiology Online, 4(3).

Zupan, Michael F1; Arata, Alan W1; Dawson, Letitia H1; Wile, Alfred L1; Payn, Tamara L2; Hannon, Megan E3 Wingate Anaerobic Test Peak Power and Anaerobic Capacity Classifications for Men and Women Intercollegiate Athletes, Journal of Strength and Conditioning Research: December 2009 - Volume 23 - Issue 9 - p 2598-2604

Please view the author's video presentation on this article at https://www.youtube.com/watch?v=kV2cr61hjcc&authuser=0

The Effect of Quality of Life on Cardiometabolic Risk Factors in Postmenopausal Women

Chavela Riotutar (Humboldt State University)

Introduction

A major health hazard is growing in our modern world that is not a communicable disease, rather a consequence of the Western lifestyle's influence. Metabolic syndrome (MetS) has become a global epidemic, resulting from increased consumption of high calorie, low fiber, processed and ultra-processed foods, compounded by a decrease in physical activity and more sedentary leisure time activities. The cluster of risk factors known as MetS is characterized by the co-occurrence of at least three of the five following criteria: waistline circumference of more than 35 inches; systolic blood pressure above 120 mmHg or diastolic blood pressure above 80 mmHg; triglyceride level above 150 mg/dL; fasting blood glucose level above 100 mg/dL; and HDL-C level below 50 mg/dL. The combination of these conditions substantially increases the risk for development of Type 2 diabetes and cardiovascular disease (CVD) (Saklayen, 2018).

Chronological aging is an additional factor as the risk for stroke doubles with every decade afterthe age of 55 (CDC, 2017). In 2017, 12.2% of adults in the United States had Type 2 diabetes and 34% had MetS (Saklayen, 2018). The economic impact of MetS is evidenced by the high usage of medical care associated with the individual components of MetS, i.e., a 40% increase inmedical care costs after the incidence of diabetes, amounting to trillions worldwide (Nichols & Moler, 2011). In a longitudinal, 9-year study by Janssen et al. (2010), menopause-related testosterone levels were implicated as a hormonal change associated with three of the five components of MetS due to the accumulation of intraabdominal fat that contains more androgen receptors. The research suggests that a decrease in estrogen levels, which promote gluteo- femoral fat accumulation, leads to the androgen-dominated hormonal milieu that increases a woman's risk for MetS. Before the age of 45, CVD is rare among women but after the age of 65,cardiometabolic risk factors accelerate until they surpass those in men, demonstrating that menopause increases the risk of CVD independent of normal aging (Ebtekar et al, 2018). The emergence of MetS indicators during menopausal transition coincides with the decline in estrogen production and may explain the higher rate of CVD in postmenopausal women (Carr,2003). Due to the hormonal changes of menopause, CVD is the primary cause of mortality forwomen in the United States, responsible for about 1 in every 5 deaths in 2017 (CDC, 2020).

The gradual process leading to menopause begins around age 45 when fertility declines as the ovaries get smaller and produce less of the hormones (estrogen and progesterone) that control the menstrual cycle (Pinkerton & Stovall, 2010). Perimenopause begins about three to five years before menopause and is characterized by irregular menstrual cycles, vasomotor symptoms, and psychological reactions (Larroy et al., 2020). Menopause is reached when a woman has no menstrual cycle for twelve months without any other cause, (e.g., illness, medication), and post menopause is the final stage, one year since the last menstrual cycle until end of life. The direct effect of estrogen deficiency on body fat distribution (central obesity), insulin action, and stiffening of the arterial wall increase the chance for a menopausal woman todevelop MetS by 60% (Carr, 2003). Identification and treatment of these indicators for MetS at an early stage provides the opportunity to prevent or postpone diabetes and CVD by making changes to daily lifestyle. Modifications to physical activity and diet may reduce future health care needs and improve overall well-being and quality of life (QOL) during menopause.

The significance of QOL as a public health concern was established by the World Health Organization (WHO) in 1949; health is not just an absence of disease, it is a state of complete physical, mental, and social well-being determined by: personal habits, social engagement, education/income, and living environment, (WHO, 2019). A healthy living environment is associated with a lower incidence of diabetes and hypertension while adverse living conditions increase the production of cortisol and the risk for CVD, (Diez Roux et al., 2016). Transitioning into menopause has been found to have a consistently negative impact on QOL, health outcomes, and increased health risks. However, research surrounding this topic fails to address several important variables associated with a woman's physiological aging process such as changes in sexual activity, alterations in sleep patterns, increased caregiving responsibilities, and severity ofchronic medical conditions that occur during menopause (Hess et al., 2012). The current life expectancy of women in the United States is 81.1 years (CDC, 2020). Accordingly, it can be saidthat women who live 81 years spend a significant part of their life in post-menopausal status.

Menopause and MetS are both associated with an unsatisfactory QOL in several studies linking social factors to health. By considering all domains that impact postmenopausal health, we can better understand how to assist this population to enhance QOL and improve health outcomes.

Research demonstrates a correlation between low QOL scores and increased risk for cardiometabolic diseases. Additionally, menopause has been correlated with lower QOL scores as well as increased risk for cardiometabolic diseases. The direct connection between these three conditions has yet to be examined. Therefore, the purpose of this study is to investigate the relationship between QOL and cardiometabolic risk indicators in postmenopausal women. The objective of this study is to assess the impact of menopause on QOL and the associated risk factors of cardiometabolic disease, specifically metabolic syndrome in postmenopausal women. Hypothesis: Postmenopausal women that express a greater burden in life (stress, financial struggle, social isolation, etc.) will show greater signs of negative cardiometabolic health outcomes. Approval for this research was granted by the Institutional Review Board (IRB registration #20-061) at Humboldt State University (HSU) in December 2020.

Method

The inclusion requirements for this study are women, female since birth, (age > 45 years) with more than twelve months since the date of last menstruation (postmenopausal). Participants must have internet access, and the ability to read and respond to the confidential online survey in English. Eighty participants will be recruited through emails to selected women's groups with instructions and a link to the survey powered by Qualtrics.com through Humboldt State University. Additionally, the researcher will create new social media pages on Facebook and Instagram that are specific to this study, then post the flyer with instructions and a link to the survey on Qualtrics.com. The QOL score is measured with 20 questions that assess the socioeconomic factors identified as determinants of health: personal habits, social engagement, education/income and living environment. The survey uses an 11-point response scale ranging from extremely dissatisfied (0) to extremely satisfied (10), resulting in a total score of 20 to 220 achieved by the sum of the responses. Additional data collection includes demographics, physical characteristics, physical activity level, sexual health, current health conditions, and medical history. Given the timing of the survey, we have included questions about changes due to the impact of sheltering in place and social distancing measures instituted in March 2020. A multiple linear regression model will be used to analyze the relationship between QOL scores, current health conditions, and sociodemographic data. The survey will be administered by the participant in their own environment without a time constraint, and responses will be recorded on the Qualtrics.com secure website. Sampling will be performed between December 2020 and March 2021. There are 35 current participants. The health history data and completed consent forms will be stored in a secure folder in the Humboldt State University Qualtrics software program for the remainder of the study and a minimum of 3 years after completion.

References

Alliance Sleep Questionnaire. (n.d.). Retrieved April 30, 2020, from https://mysleep.stanford.edu/

Arnot, M., & Mace, R. (2020). Sexual frequency is associated with age of natural menopause: Results from the Study of Women's Health Across the Nation. *Royal Society Open Science*, *7*(1), 191020. https://doi.org/10.1098/rsos.191020

Carr, M. C. (2003). The Emergence of the Metabolic Syndrome with Menopause. *The Journal of Clinical Endocrinology & Metabolism*, *88*(6), 2404–2411. https://doi.org/10.1210/jc.2003-030242

CDC. (2020a, January 31). *Women and Heart Disease*. Centers for Disease Control and Prevention. https://www.cdc.gov/heartdisease/women.htm

Diez Roux, A. V., Mujahid, M. S., Hirsch, J. A., Moore, K., & Moore, L. V. (2016). The impact of neighborhoods on cardiovascular risk: The MESA Neighborhood Study. *Global Heart*, *11*(3), 353–363. https://doi.org/10.1016/j.gheart.2016.08.002

Ebtekar, F., Dalvand, S., & Gheshlagh, R. G. (2018). The prevalence of metabolic syndrome in postmenopausal women: A systematic review and meta-analysis in Iran. *Diabetes & Metabolic Syndrome: Clinical Research & Reviews*, *12*(6), 955–960. https://doi.org/10.1016/j.dsx.2018.06.002

Hess, R., Thurston, R. C., Hays, R. D., Chang, C.-C. H., Dillon, S. N., Ness, R. B., Bryce, C. L., Kapoor, W. N., & Matthews, K. A. (2012). The impact of menopause on health-related quality of life: Results from the STRIDE longitudinal study. *Quality of Life Research*, *21*(3), 535–544. https://doi.org/10.1007/s11136-011-9959-7

Janssen, I., Powell, L. H., Crawford, S., Lasley, B., & Sutton-Tyrrell, K. (2008). Menopause and the metabolic syndrome: The Study of Women's Health Across the Nation. *Archives of Internal Medicine*, *168*(14), 1568–1575. https://doi.org/10.1001/archinte.168.14.1568

Kim, H., & Cho, Y. (2020). Factors Associated with Metabolic Syndrome Among Middle-Aged Women in Their 50s: Based on National Health Screening Data. *International Journal of Environmental Research and Public Health*, *17*(9), 3008. https://doi.org/10.3390/ijerph17093008

Larroy, C., Marin Martin, C., Lopez-Picado, A., & Fernández Arias, I. (2020). The impact of perimenopausal symptomatology, sociodemographic status and knowledge of menopause on women's quality of life. *Archives of Gynecology and Obstetrics*, *301*(4), 1061–1068. https://doi.org/10.1007/s00404-019-05238-1

Menopause Health Questionnaire. (n.d.). Retrieved January 27, 2021, from https://www.menopause.org/publications/clinical-practice-materials/menopause-health-questionnaire

Nichols, G. A., & Moler, E. J. (2011). Metabolic Syndrome Components Are Associated with Future Medical Costs Independent of Cardiovascular Hospitalization and Incident Diabetes. *Metabolic Syndrome and Related Disorders*, *9*(2), 127–133. https://doi.org/10.1089/met.2010.0105

Paley, C. A., & Johnson, M. I. (2018). Abdominal obesity and metabolic syndrome: Exercise as medicine? *BMC Sports Science, Medicine and Rehabilitation*, *10*. https://doi.org/10.1186/s13102-018-0097-1

Perceived Quality of Life Scale (PQOL)—Seattle Quality of Life Group—Measure and Improve Health Disparities in Children, Adolescents, Adults from Stigmatized Populations. (n.d.). Retrieved December 13, 2019, from http://depts.washington.edu/seaqol/PQOL

Pinkerton, J. V., & Stovall, D. W. (2010). Reproductive aging, menopause, and health outcomes: Pinkerton & Stovall. *Annals of the New York Academy of Sciences*, *1204*(1), 169–178. https://doi.org/10.1111/j.1749-6632.2010.05526.x

Saklayen, M. G. (2018). The Global Epidemic of the Metabolic Syndrome. *Current Hypertension Reports*, *20*(2). https://doi.org/10.1007/s11906-018-0812-z

Stefanska, A., Bergmann, K., & Sypniewska, G. (2015). Chapter One - Metabolic Syndrome and Menopause: Pathophysiology, Clinical and Diagnostic Significance. In G. S. Makowski (Ed.), *Advances in Clinical Chemistry* (Vol. 72, pp. 1–75). Elsevier. https://doi.org/10.1016/bs.acc.2015.07.001

Taebi, M., Abdolahian, S., Ozgoli, G., Ebadi, A., & Kariman, N. (2018). Strategies to improve menopausal quality of life: A systematic review. *Journal of Education and Health Promotion*, *7*(1), 93. https://doi.org/10.4103/jehp.jehp_137_17

Vetter, M. L., Wadden, T. A., Lavenberg, J., Moore, R. H., Volger, S., Perez, J. L., Sarwer, D. B., & Tsai, A. G. (2011). Relation of health-related quality of life to metabolic syndrome, obesity, depression and comorbid illnesses. *International Journal of Obesity*, *35*(8), 1087–1094. https://doi.org/10.1038/ijo.2010.230

Vincent, G. E., Jay, S. M., Sargent, C., Vandelanotte, C., Ridgers, N. D., & Ferguson, S. A. (2017). Improving Cardiometabolic Health with Diet, Physical Activity, and Breaking Up Sitting: What about Sleep? *Frontiers in Physiology*, *8*, 865. https://doi.org/10.3389/fphys.2017.00865

WHO | WHOQOL: Measuring Quality of Life. (n.d.). WHO. Retrieved February 13, 2020, from https://www.who.int/healthinfo/survey/whoqol-qualityoflife/en/

Please view the author's video presentation on this article at https://digitalcommons.humboldt.edu/csucompetitionvideos/1/

Made in United States
Troutdale, OR
11/03/2026

56844676R00064